High Performance Computing on Vector Systems 2008

Michael Resch · Sabine Roller · Katharina Benkert ·
Martin Galle · Wolfgang Bez · Hiroaki Kobayashi ·
Toshio Hirayama
Editors

High Performance Computing on Vector Systems 2008

Springer

Michael Resch
Sabine Roller
Katharina Benkert
Höchstleistungsrechenzentrum
Stuttgart (HLRS)
Universität Stuttgart
Nobelstraße 19
70569 Stuttgart
Germany
resch@hlrs.de
roller@hlrs.de
benkert@hlrs.de

Martin Galle
Wolfgang Bez
NEC Deutschland GmbH
Hansaallee 101
40549 Düsseldorf
Germany
mgalle@hpce.nec.com
wbez@hpce.nec.com

Hiroaki Kobayashi
Cyberscience Center
Tohoku University
6-3 Aramaki-Aza-Aoba
Sendai, 980-8578
Japan
koba@isc.tohoku.ac.jp

Toshio Hirayama
Center for computational science and e-systems
Japan Atomic Energy Agency
Sumitomo fudosan Ueno Bldg. No. 8
6-9-3 Higashi-Ueno Taito-ku
Tokyo, 110-0015
Japan
hirayama.toshio@jaea.go.jp

Front cover figure: Simulation of the UV curing process in automotive coating with multiple ultraviolet lamps. Picture due to IFF, University of Stuttgart, Germany and BMW Group.

ISBN 978-3-642-09941-0 e-ISBN 978-3-540-85869-0

DOI 10.1007/978-3-540-85869-0

Mathematics Subject Classification (2000): 68Wxx, 68W10 , 68U20, 76-XX, 86A05, 86A10, 70Fxx

Cover design: WMXDesign, Heidelberg

Printed on acid-free paper

9 8 7 6 5 4 3 2 1

springer.com

Preface

This book covers the results obtained in the Teraflop Workbench project during a four years period from 2004 to 2008. The Teraflop Workbench project is a collaboration between the High Performance Computing Center Stuttgart (HLRS) and NEC Deutschland GmbH (NEC-HPCE) to support users to achieve their research goals using high performance computing.

The Teraflop Workbench supports users of the HLRS systems to enable and facilitate leading edge scientific research. This is achieved by optimizing their codes and improving the process workflow which results from the integration of different modules into a "hybrid vector system". The assessment and demonstration of industrial relevance is another goal of the cooperation.

The Teraflop Workbench project consists of numerous individual codes, grouped together by application area and developed and maintained by researchers or commercial organizations. Within the project, several of the codes have shown the ability to reach beyond the TFlop/s threshold of sustained performance. This created the possibility for new science and a deeper understanding of the underlying physics. The papers in this book demonstrate the value of the project for different scientific areas.

The work in the Teraflop Workbench project gave us insight into the applications and requirements for current and future HPC systems. We observed the emergence of multi-scale and multi-physics applications, the increase in interdisciplinary work and the growing tendency to use today's stand-alone application codes as modules in prospective, more complex coupled simulations. At the same time, we noticed the current lack of support for those applications. Our goal is to offer an environment to our users that allows them to concentrate on their area of expertise without spending too much time on computer science itself.

We would like to thank all the contributors of this book and of the Teraflop Workbench project in general.

Stuttgart, July 2008

Sabine P. Roller
Michael M. Resch

Contents

I Future Architectures

II Multiscale and Multiphysics Simulations

III Grid Computing & Data Analysis

I
Future Architectures

First Experiences with NEC SX-9

Hiroaki Kobayashi, Ryusuke Egawa, Hiroyuki Takizawa, Koki Okabe, Akihiko Musa, Takashi Soga, and Yoichi Shimomura

Abstract This paper presents the new supercomputer system NEC SX-9 that has been installed at Tohoku University in March 2008. The performance of the system is evaluated by using six real application codes. The experimental results indicate that the SX-9 system achieves a speedup of up to 7 compared to our previous NEC SX-7 system for the single-CPU sustained performance. In addition, the paper examines the effects of an on-chip vector cache named ADB on the performance, and confirms performance increases between 20 and 70% by selective caching on ADB.

1 Introduction

The Cyberscience Center (formerly Information Synergy Center) is one of seven national supercomputer centers in Japan. Since 1986, we have installed the latest vector supercomputer system at the time. Fig. 1 shows the improvements in performance and memory capacity of the five generations of supercomputer systems at Tohoku University. As shown, the latest system was installed every four to five years. Each achieves a performance improvement of 8 times or more in both Flop/s rate and memory capacity. In March 2008, we have installed a new supercomputer system named NEC SX-9 (see Fig. 2), which employs the first single 100+ GFlop/s

Hiroaki Kobayashi · Ryusuke Egawa · Hiroyuki Takizawa · Koki Okabe
Tohoku University, Sendai 980-8578, Japan,
e-mail: {koba, egawa, tacky, okabe}@isc.tohoku.ac.jp

Akihiko Musa
Tohoku University, Sendai 980-8578, Japan
NEC Corporation, Tokyo 108-8001, Japan, e-mail: musa@sc.isc.tohoku.ac.jp

Takashi Soga
NEC System Technologies, Osaka 540-8551, Japan, e-mail: soga-txa@necst.nec.co.jp

Yoichi Shimomura
NEC Software Tohoku, Sendai 980-0811, Japan, e-mail: y-shimomura@wx.jp.nec.com

vector processor. This paper gives an overview of our SX-9 system, the first machine of this kind delivered by NEC. We report its performance on real science and engineering applications, especially in terms of sustained performance of the 100+ Gflop/s vector processor and the effects of ADB, which is a newly designed on-chip cache for the SX vector processor.

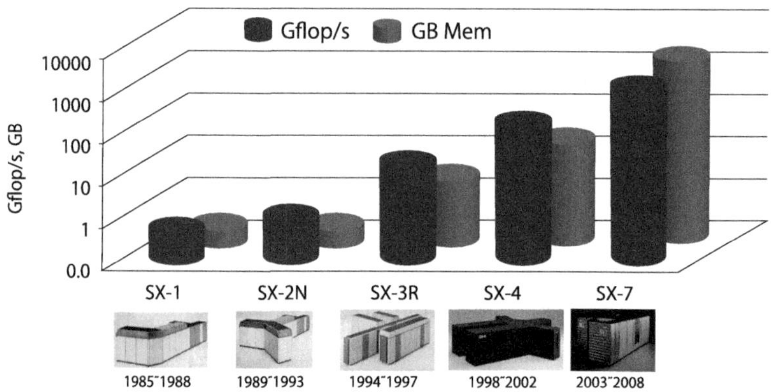

Fig. 1 Five Generations of NEC SX systems at Tohoku University

Fig. 2 NEC SX-9

2 System Overview

Fig. 3 shows the organization of our supercomputer systems. The SX-9 system consists of 16 nodes, each having 16 vector processors sharing a large memory space of 1 TB for shared memory processing (SMP). The 16 nodes are interconnected via a custom designed high-bandwidth crossbar network named IXS (Interconnect Crossbar Switch) at 256 Gbits (bi-directional). As each processor has a peak performance of 102.4 GFlop/s, the total performance of the system is 26.2 TFlop/s.

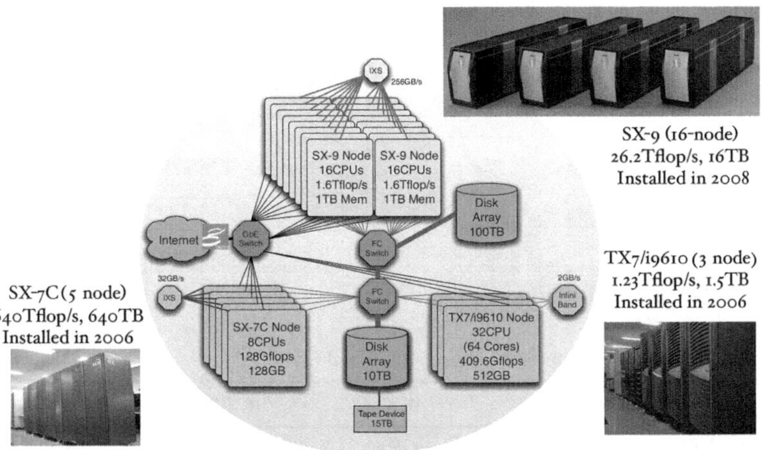

Fig. 3 Supercomputers at Tohoku University

Fig. 4 shows the architecture of the vector processor. The vector processor of SX-9 is designed in 65 nm technology and its operation frequency is 3.2 GHz. The SX-9 processor doubles both vector pipelines and vector units, and triples the clock frequency compared with the SX-7 vector processor (our previous system), resulting in a speedup of 11.6 for the single processor performance. The SX-9 processor also employs a newly designed 256 KB on-chip cache named ADB (Assigned Data Buffer) to support the vector load-store unit. By keeping vector data with locality in the ADB, a 4 Byte/Flop rate is guaranteed to effectively drive most of the vector pipes on the chip. Table 1 compares the performance of the SX-9 and the SX-7 system. The SX-9 offers a balanced combination of high performance vector CPUs and a larger shared main memory, which are connected with a higher memory bandwidth compared to the SX-7 system. This provides a user-friendly high-performance computing environment, in which users enjoy the potential of the system with less efforts.

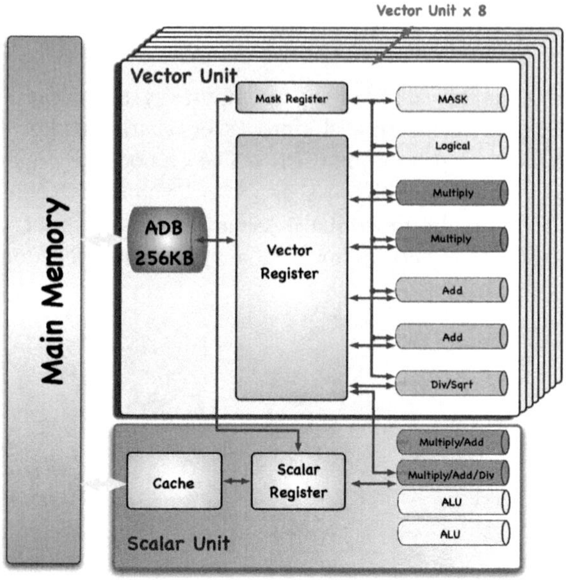

Fig. 4 SX-9 Processor Architecture

Table 1 Performance Comparison between SX-9 and SX-7

		SX-7 in 2003	SX-9 in 2008	Factor of Increase
per CPU	Freq.	1.1 GHz	3.2 GHz	2.9
	Vec. Perf.	8.83 GFlop/s	102.4 GFlop/s	11.6
	Mem. BW	35.3 GB/s	256 GB/s	7.3
per SMP node	Vec. Perf.	282 GFlop/s	1.6 TFlop/s	5.8
	Mem. Cap.	256 GB	1 TB	4.0
	Mem. BW	1.13 TB/s	4 TB	3.5
	Mem. Banks	16000	32000	2.0
	IXS BW	32 GB/s[a]	256 GB/s	8.0
per System	Total Perf.	2.1 TFlop/s	26.2 TFlop/s	12.5
	Total Mem.	2 TB	16 TB	8.0

[a] SX-7C(8)

3 Performance Evaluation

In this section, we examine the benchmark results of the SX-9 system. Table 2 shows six real applications that we have used for the performance evaluation. These applications are designed and developed by our users in the fields of science and engineering. In the performance evaluation, we are especially interested in the sustained performance of the 102.4 GFlop/s vector processor and the effects of the ADB on supporting vector load store operations.

Fig. 5 shows the sustained single CPU performance of the six benchmark programs for five NEC supercomputer systems, which are SX-9 (102.4 GFlop/s peak performance), SX-7 (8.8 GFlop/s peak performance), SX-8 (16 GFlop/s peak performance), SX-8R (35.2 GFlop/s peak performance) and TX7/i9610 ItaniumII (6.4 GFlop/s peak performance). The benchmark programs are compiled by the NEC Fortran compiler with automatic vectorization and automatic parallelization for the SX and TX7 systems. The SX-9 performs significantly better compared to the other systems, especially for the Antenna code. As the Antenna code has a high number of arithmetic operations per vector load/store, the enhanced pipelines and vector units as well as the higher clock frequency of the SX-9 linearly improve the sustained performance. The Plasma simulation requires a lot of list vector accesses and the gather/scatter operations put a large pressure on the load/store unit. The resulting sustained performance of the SX systems is not so high due to the bottleneck in memory accesses, but it is still impressive compared to ItaniumII.

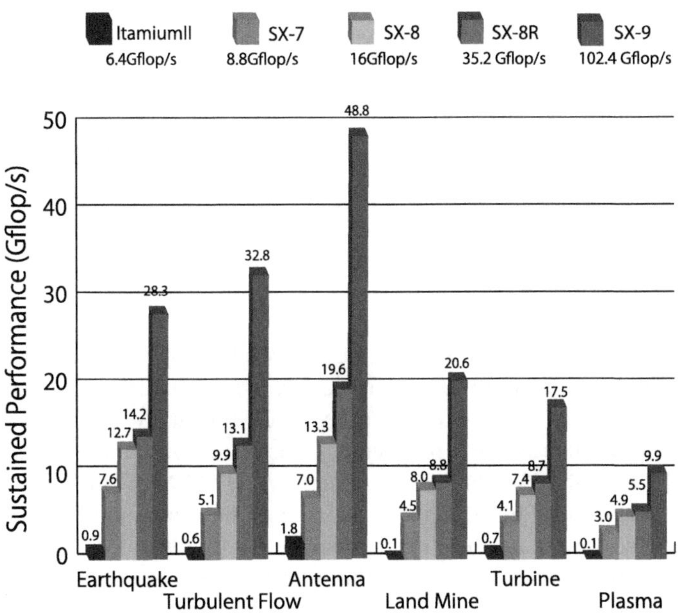

Fig. 5 Single CPU performance of six benchmarks

Fig. 6 depicts the direct comparison in the sustained performance between the SX-9 single processor and ItaniumII. As Fig. 6a shows, the SX-9 has a maximum speedup of 158 compared with a single ItaniumII CPU performance. Its average performance across all benchmarks is a factor of 67 higher as the ItaniumII even though the peak performance ratio between the two systems is only 16. Even in the case of a system with 16 ItaniumII CPUs, whose peak performance is the same as

Table 2 Benchmark Programs

Earthquake [1]	Simulation of seismic slow slip model.
Turbulent flow [7]	Direct numerical simulation of turbulent channel flow.
Antenna [5]	FDTD simulation of lens antenna using Fourier transform.
Land Mine [3]	FDTD simulation of array antenna ground penetrating radar for land mine detection.
Turbine [4]	Direct numerical simulation of unsteady flow through turbine channels for hydroelectric generators.
Plasma [2]	Simulation of upper hybrid wave in plasma using Lax-Wendroff method.

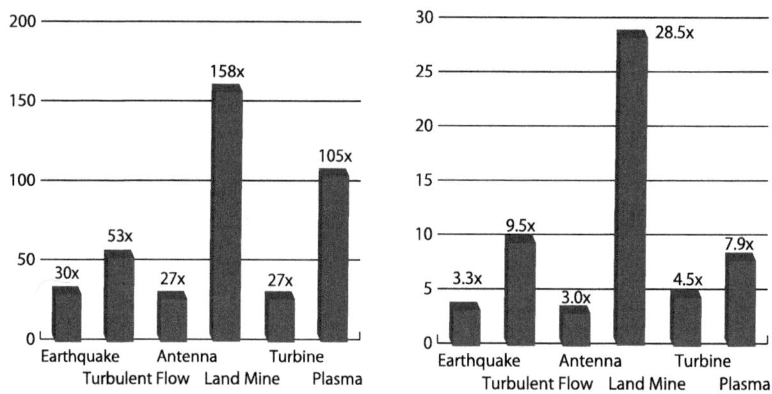

a vs. one ItaniumII CPU **b** vs. 16 ItaniumII CPUs

Fig. 6 SX-9 singe CPU performance vs. ItaniumII

for one CPU SX-9, the vector CPU achieves speedups of 3.0 up to 28.5 as shown in Fig. 6b. This significant difference in performance between SX-9 and ItaniumII is due to their memory bandwidths. For the Land Mine application, more than 95% of total execution time is spent for memory operations on ItaniumII because of its low cache hit rate of 70%. On the other hand, on NEC SX-9, only 25% of the total processing time results from memory access and the remaining 75% is available for arithmetic operations. The higher memory bandwidth of the SX-9 decreases the portion of memory processing in the total time and contributes to the higher sustained performance compared to the scalar system.

In the following, we examine the effects of the 256 KB ADB on the execution times of the applications. Fig. 7a shows one of the most time-consuming kernels of the Land Mine application. The Land Mine application is based on the FDTD method and has many data references with high locality. In this kernel, we selectively cache array `H_x` with the directive `!cdir ON_ADB(H_x)` as shown in Fig. 7b. By selective caching, vector loads of `H_x(i,j-1,k)` bring data into cache. This data will be reused subsequently for vector loads of `H_x(i,j,k)` and causes high cache hit rates. Fig. 8a shows the performance of SX-9 with ADB normalized by that without ADB. A performance improvement of 20% is obtained by ADB with selective caching. The Earthquake application also profits from the usage of the ADB. Fig. 9 displays one of the computationally most expensive kernels. If we selectively cache the loop-independent array `wary(j)` by pre-fetching it before loop, a performance improvement of 70% can be obtained as shown in Fig. 8b.

```
      DO 10 k=0,Nz
      DO 10 i=0,Nx
      DO 10 j=0,Ny
         E_x(i,j,k) = C_x_a(i,j,k)*E_x(i,j,k)
 &          + C_x_b(i,j,k) * ( (H_z(i,j,k)-H_z(i,j-1,k))/dy
 &          -(H_y(i,j,k)-H_y(i,j,k-1))/dz -E_x_Current(i,j,k) )
         E_y(i,j,k) = C_y_a(i,j,k)*E_y(i,j,k)
 &          + C_y_b(i,j,k) * ( (H_x(i,j,k)-H_x(i,j,k-1))/dz
 &          -(H_z(i,j,k)-H_z(i-1,j,k))/dx -E_y_Current(i,j,k) )
         E_z(i,j,k) = C_z_a(i,j,k)*E_z(i,j,k)
 &          + C_z_b(i,j,k) * ( (H_y(i,j,k)-H_y(i-1,j,k))/dx
 &          -(H_x(i,j,k)-H_x(i,j-1,k))/dy -E_z_Current(i,j,k) )
 10   CONTINUE
```

a Original code

```
      DO 10 k=0,Nz
      DO 10 i=0,Nx
!cdir ON_ADB(H_x)
      DO 10 j=0,Ny
         E_x(i,j,k) = C_x_a(i,j,k)*E_x(i,j,k)
 &          + C_x_b(i,j,k) * ( (H_z(i,j,k)-H_z(i,j-1,k))/dy
 &          -(H_y(i,j,k)-H_y(i,j,k-1))/dz -E_x_Current(i,j,k) )
         E_y(i,j,k) = C_y_a(i,j,k)*E_y(i,j,k)
 &          + C_y_b(i,j,k) * ( (H_x(i,j,k)-H_x(i,j,k-1))/dz
 &          -(H_z(i,j,k)-H_z(i-1,j,k))/dx -E_y_Current(i,j,k) )
         E_z(i,j,k) = C_z_a(i,j,k)*E_z(i,j,k)
 &          + C_z_b(i,j,k) * ( (H_y(i,j,k)-H_y(i-1,j,k))/dx
 &          -(H_x(i,j,k)-H_x(i,j-1,k))/dy -E_z_Current(i,j,k) )
 10   CONTINUE
```

b Code with the directive for selective caching

Fig. 7 Kernel of the Land Mine benchmark

4 Conclusions

In this paper we gave an overview of the new NEC SX-9 supercomputer installed at Tohoku University and examined its performance when executing science and engineering applications. Our system is the first one equipped delivered by NEC with 100+ GFlop/s vector chips. Various new technologies such as ADB (Assigned Data Buffer) are incorporated into it. The experimental results show remarkable

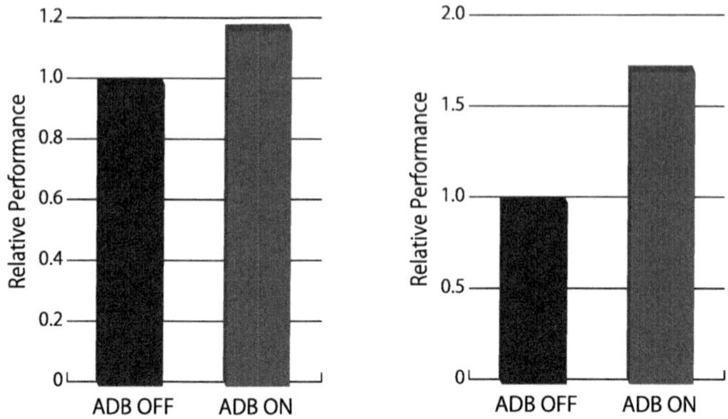

a Land Mine **b** Earthquake

Fig. 8 Effects of ADB

```
do i=1,ncells
do j=1,ncells
     wf_dip(i)=wf_dip(i)+gd_dip(j,i)*wary(j)
end do
end do
```

Fig. 9 Kernel of Earthquake

improvements for the considered applications. In this paper, only the single CPU performance of SX-9 has been discussed, but the scalable performance of our multi-node SX-9 system on a CFD MPI code has already been reported [6]. In our opinion, NEC SX-9 has the ability to become a key supercomputing platform.

Acknowledgements This work was part of a collaboration between Tohoku University and NEC, and many colleagues contributed to this project. We would also like to thank Professors Akira Hasegawa, Kunio Sawaya, Motoyuki Sato, Satoru Yamamoto, Yasuhiro Sasao, Associate professor Masahide Iizima of Tohoku University, Dr. Takahiro Tsukahara of Tokyo University of Science and Dr. Keisuke Ariyoshi of JAMSTEC for providing the benchmark codes.

References

1. Ariyoshi, K., Matsuzawa, T., Hasegawa, A.: The key frictional parameters controlling spatial variation in the speed of postseismic slop propagation on a subduction plate boundary. Earth and Planetary Science Letters **256**, 136–146 (2007)
2. Katoh, Y., Iizima, M.: A computer simulation study on the mode conversion process from slow X-mode to fast X-mode by the tunneling effect. In: Earth, Planets and Space, pp. e53–e56 (2006)
3. Kobayashi, T., et al.: FDTD simulation on array antenna SAR-GPR for land mine detection. In: Proceedings of SSR2003, pp. 279–283 (2003)

4. Sasao, Y., et al.: Numerical prediction of unsteady flows through turbine stator-rotor channels with condensation. In: Proceedings of ASME Fluids Engineering Summer Conference (2005)
5. Takagi, Y., et al.: Study of high gain and broadband antipodal fermi antenna with corrugation. In: Proceedings of 2004 International Symposium on Antennas and Propagation, vol. 1, pp. 69–72 (2007)
6. Takahashi, S., et al.: Large scaled computation of incompressible flows on cartesian mesh using a vector-parallel supercomputer. In: Parallel CFD 2008 (2008)
7. Tsukahara, T., et al.: Evolution of material line in turbulent channel flow. In: Proceedings of the Fifth Int. Symp. on Turbulence and Shear Flow Phenomena, pp. 549–554 (2007)

Scalable Computing in a Hybrid System Architecture

Wilfried Oed

Abstract The largest performance gains in current supercomputers come from increased parallelism on practically all levels. Within a processor, there are multiple functional units as well as multiple cores. In order to reach Petaflop scale, well over 100000 processor cores need to be employed in such a system. This trend is expected to continue for the next couple of years, thus the biggest challenge is making this computational power available to a wider spectrum of "real" applications, like fluid dynamics or complex weather models. In order to increase single processor performance, various concepts of hybrid computing are under investigation. This paper covers technology trends, an overview of the current Cray $XT5_h$ hybrid supercomputer, and provides an outlook to future Cray MPP architectures.

1 Introduction

About 30 years ago, the term "supercomputer" was first defined with the advent of the Cray-1 displayed in Fig. 1. While mostly renowned for its innovative vector architecture, an often forgotten aspect is that it also was the fastest scalar machine on the planet at that time. As a pleasant property practically any application ran faster than on any other system, and those spending time and effort to vectorize were rewarded with speedups unheard of before. High user productivity also was attributable to a simple Operating System, an auto-vectorizing compiler with support through directives, and the easy to use yet effective *Flowtrace* performance analysis tool.

Finally, the Cray-1 boasted dense packaging accompanied by innovative liquid cooling.

From this outset, we'll explore in the following what this means in today's HPC world.

Wilfried Oed
Cray Computer GmbH, Waldhofer Str. 102, 69123 Heidelberg, Germany, e-mail: wko@cray.com

Fig. 1 Cray-1

2 The Need for Scalability

Whilst the TOP 500 list [1] has little to do with actual application performance, it nonetheless reflects architectural and technological trends. Using the Cray-1 again as a starting point, this system achieved about 150 MFlop/s on the LINPACK benchmark, current microprocessors reach about 50 times as much on a single processor core. The target performance for the most powerful systems in the world just has become 1 PFlop/s, which is more than 6.5 million times the Cray-1 performance! Since the processor performance "only" has increased by a factor of 50, this implies that more than 100000 Cray-1 type processors comprise such a system!

And indeed even the average number of processors (or processor cores) employed in the 10 most powerful systems of the TOP 500 is close to 100000 processors. As shown in Fig. 2, this trend really took off about five years ago and continues to rise.

Having built such large systems, the real challenge is making this computational power available to a wider spectrum of "real" applications, like fluid dynamics or complex weather models. Aspects like communication, load imbalance, and I/O now come into play. Hardware and system software have to operate in a reliable and consistent way. Any kind of disturbances, commonly referred to as *jitter* are inhibiting scalability. The impact of OS jitter has been observed in an important study by Petrini, Kerbyson, and Pakin who presented the results at the SC03 conference [5].

Having scalability in mind, Cray MPP systems run Compute Node Linux (CNL), a Linux based lightweight kernel operating system on their compute nodes. CNL includes a runtime environment based on the SUSE SLES distribution and provides outstanding performance and scaling characteristics. Current Cray XT4 systems typically are used for capability computing, with actual workloads utilizing more than 15000 processor cores.

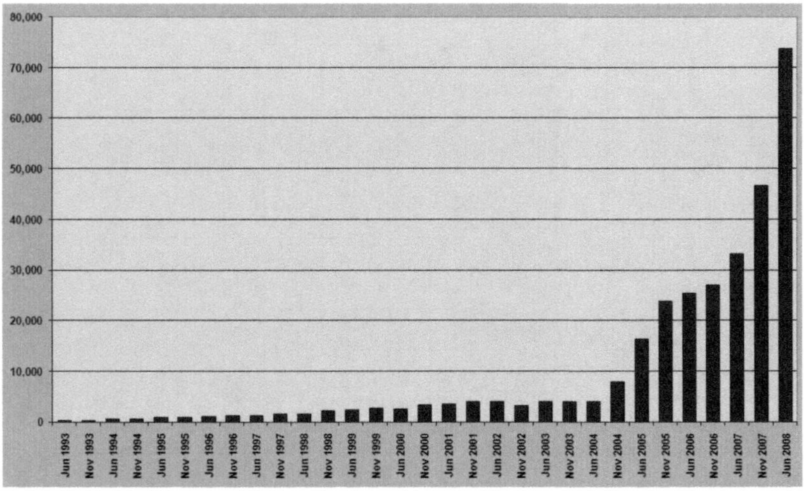

Fig. 2 Evolution of average number of processors of the 10 most powerful systems of the Top 500

3 Processor Trends

Ever since the Cray-1, vector processors continued to keep their lead in terms of absolute performance. However, they failed to maintain their "no matter what" advantage, i.e. being the fastest system on any kind of application. Driven by mass market requirements, microprocessor speeds began to overtake vector processors by far on non-vector code. In addition, the much lower price of a microprocessor enabled massively parallel systems. Starting in the mid-90s, applications began to be split into vector and non-vector, with non-vector becoming more dominant over time. The latter may sound surprising, but as supercomputers became more powerful, applications could afford to become more complex that way providing more insight to science. Aside from finer resolutions, this also meant more physics and chemistry parameters, which typically is not amenable to vector processing.

Figure 3 shows the single processor performance achieved on LINPACK [3, 4]. In order to stay in the list, these clusters typically employ state of the art microprocessors.

As expected vector processors are well above average, for MPP systems however two distinct approaches can be seen: One is to utilize an extremely large number of fairly weak processors, the other is to follow the processor roadmap. Cray has deliberately chosen to utilize the fastest commodity microprocessor for their MPP systems, as it is amenable to a wider spectrum of applications. This has been true for the Cray T3E in the past as well as the current Cray XT4 and Cray XT5.

Quite recently, a shift in microprocessor design can be observed. For quite some time, microprocessors made huge strides by faster clock speeds which then was stopped at the 2 to 3 GHz range. This was not so much for having reached funda-

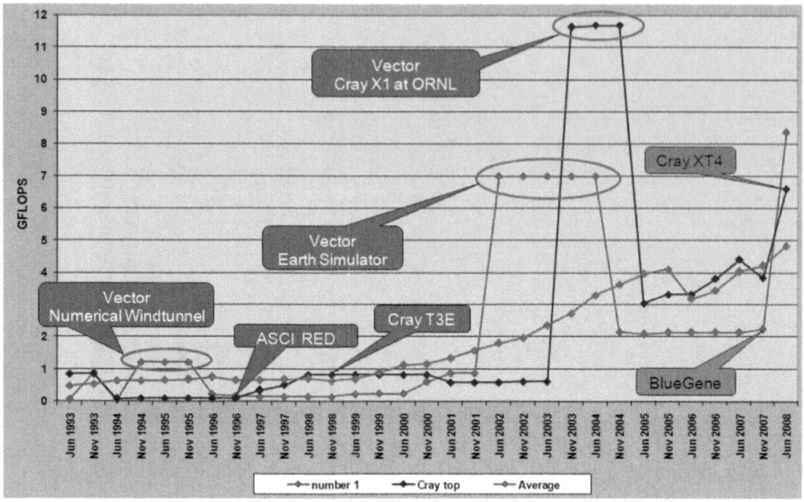

Fig. 3 Evolution of the single processor performance on LINPACK

mental technical limits, rather power and cooling began to offset speed advantages, particular for mass market requirements.

Additional performance gains now come from increased parallelism within the processor itself. This for one are multiple functional units as well as multiple cores within a processor.

The most promising approach for utilizing this form of parallelism is a hybrid programming paradigm with MPI message passing across nodes and OpenMP within a node. The major benefits are a more efficient use of the interconnection network by fewer and larger messages and providing load balancing within a node. This concept already has successfully been demonstrated on full applications to well over 15000 processor cores on the Cray XT4 supercomputer on current Quad-Core AMD Opteron processors.

The term *hybrid computing* encompasses also the combination of different processor architectures to better match application characteristics. A much debated approach is the use of specialized floating point accelerators. While this works well for Linpack, since it fully executes on the accelerator card, it is completely unclear how full applications can take advantage of this concept. The difficulty with accelerator cards is the huge overhead involved when switching and moving data between the two types of architectures.

Cray therefore takes a different approach to match application characteristics. The first step is the hybrid Cray $XT5_h$ system where two distinct architectures are tightly coupled within a common infrastructure. This is available now and briefly described in Sect. 4.

The next step will provide the integration of different architectures into a common address space. This concept is currently being developed under the project name *Cascade*. A conceptual description is provided in Sect. 5.

4 The Cray XT5$_h$ Hybrid Supercomputer

Recognizing that no single architecture is ideally suited for all types of applications, Cray introduced the Cray XT5$_h$ [6], the industry's first hybrid supercomputer. It can integrate multiple processor architectures into a single system and accelerate high performance computing workflows. Figure 4 exemplifies this approach for coupled climate models.

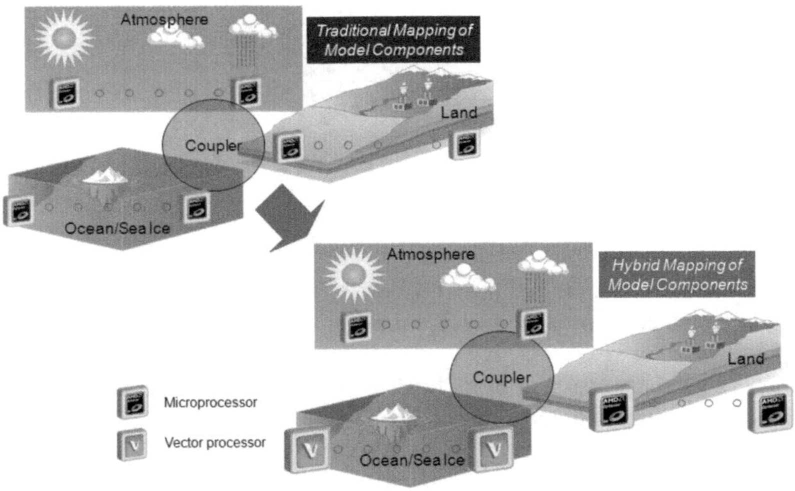

Fig. 4 Hybrid mapping of a coupled climate model

Contrary to accelerators, each model fully executes on the optimal platform. As such, the distinction is on a much coarser level without a loss in performance from switching between the two architectures.

The major building blocks of the Cray XT5$_h$ system are scalar processor blades as used in the Cray XT4 and Cray XT5 MPP systems, Cray X2 vector blades, and Service (SIO) blades for Login, Networking, I/O and general System Services (see Fig. 5).

A scalar Cray XT5 blade includes four compute nodes in a small footprint — up to 300 GFlop/s per blade. Each compute node is composed of two AMD Opteron^{TM} processors (dual or quad core), each coupled with its own memory and dedicated Cray SeaStar2+^{TM} communication Application-Specific Integrated Circuit (ASIC). Cray XT4 blades only utilize one AMD processor per node, thus providing a different computation to communication ratio.

The Cray XT4 and Cray XT5 blades incorporate a high-bandwidth, low-latency interconnect composed of Cray SeaStar2+^{TM} chips and high-speed links based on HyperTransport and proprietary protocols. The interconnect directly connects all scalar compute nodes in a 3D torus topology, eliminating the cost and complexity of external switches. This improves reliability and allows the Cray XT5 system to

Fig. 5 Layout of the integrated hybrid supercomputer CRAY $XT5_h$

economically scale to tens of thousands of nodes. As the backbone of the Cray XT5 system, the interconnect carries all message passing traffic as well as all I/O traffic to the global file system.

The Cray X2 vector processing blade holds two Cray X2 compute nodes and is the extreme high bandwidth processing component of the Cray $XT5_h$. Based on four powerful vector processors, each Cray X2 node is capable of more than 100 GFlop/s peak performance in 64-bit mode, twice as much in 32-bit mode. The four vector CPUs are implemented as a 4-way SMP, with each CPU having uniform memory access to the local node memory. Systems can be scaled to 32000 processors using an extremely high-bandwidth, low-latency interconnect in a fat tree topology.

The Cray X2 interconnect interfaces directly with the Cray SeaStar2+^{TM} network on the Cray XT5 system. This supports common application development, I/O and storage, system administration, user login, and hybrid computing.

The Cray $XT5_h$ supercomputer provides an established and complete software development environment to harness the full power of the systems hybrid supercomputing capabilities. This environment includes proven and powerful vector and scalar compilers, scientific and math libraries, debuggers, and performance analysis tools.

The Cray $XT5_h$ system runs a Linux environment, providing a single point of login across all system resources or blades. A common Application Level Placement Scheduler (ALPS) allows for affinity aware application scheduling across the system's scalar and vector blades. The scheduler also fully integrates with workload management applications including PBS Pro and LSF, guaranteeing resource availability for batch and interactive requests.

A Lustre® file system provides shared file access of common input and output files to all applications and computational resources in the workflow. Applications can easily and immediately access the output of one application from one resource as the input to another application within the workflow. Figure 6 shows the globally accessible filesystem.

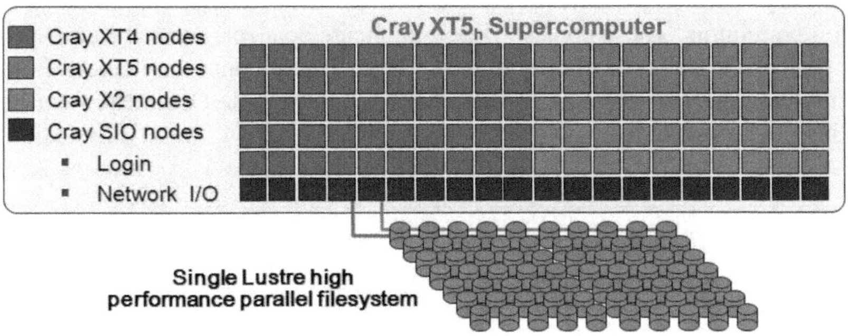

Fig. 6 Layout of the globally acessible filesystem of the CRAY $XT5_h$

The Cray $XT5_h$ system supports Parallel Global Address Space (PGAS) languages within each architectural domain. Socket-based communication between applications also facilitates advanced hybrid application development across domains.

For vector processing, Co-Array Fortran (CAF) and Unified Parallel C (UPC) support in the compiler is closely integrated with the optimizer. This allows the emitted code to take full advantage of the vector processing's memory architecture for low latency communication.

5 The Cray Cascade Program

The U.S. Defense Advanced Research Projects Agency (DARPA) introduced the High Productivity Computing Systems (HPCS) program to facilitate development of the next generation of supercomputers for the national security, research, and industrial user communities. Cray's HPCS development program, code-named *Cascade* [2], will produce a system with exceptional programmability, portability and robustness — capable of scaling to unprecedented levels of sustained performance on real applications. Under the Cascade program, Cray will develop a new hybrid system architecture that combines multiple processor technologies, a new high-performance network and an innovative adaptive software layer into a single integrated system. Designed to efficiently scale to large numbers of processors, the system will maximize productivity and performance by automatically matching the most effective processor technology to each application.

The following diagram 7 shows the principle architectural concept of tightly integrating different compute *locales*. A locale is a computational entity providing specific functionality for vectorizable, scalar or other types of applications. SIO locales provide access as well as high bandwidth networking and I/O to a global filesystem.

A high radix interconnection network provides extremely high bandwidth and low latency to all locales within the system. Atomic memory operations (AMO) within the network provide the basis for a globally addressable memory space. The atomic memory operations enable programmers to efficiently utilize simplified communication paradigms. These consist of new language concepts as well as the efficient support of Co-Array Fortran and UPC. High performance implementations of proven communication mechanisms like MPI and intra-node OpenMP of course will continue to be supported.

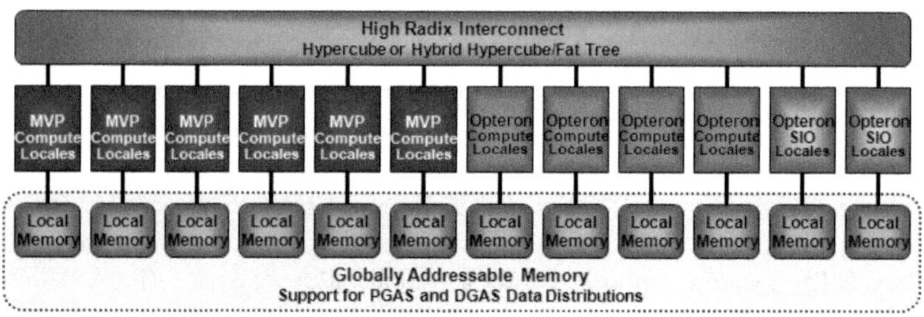

Fig. 7 Highly integrated locales in the CRAY $XT5_h$

6 Conclusion

Taking capability computing to new heights requires scaling applications to very large number of processors. Cray MPP supercomputers as well as the accompanying software are designed to scale. The endeavor for improvement is targeted at closely integrating hybrid paradigms to reduce or even eliminate associated overhead. The next generation Cascade program incorporates the hybrid computing paradigms into a single system. Thus, the two distinct domains of the current hybrid Cray $XT5_h$ supercomputer will then be integrated into a single system in a globally addressable memory space. Lastly, it should be mentioned, that innovative power and cooling options have been developed for these systems and will be made public at the SC08 conference.

References

1. Top 500 supercomputing sites. http://www.top500.org/
2. In Cray's "Cascade", the computer will adapt to the codes. HPCwire (2006). http://www.hpcwire.com/features/17883334.html
3. Dongarra, J.J., Bunch, J.R., Moler, G.B., Stewart, G.W.: LINPACK Users' Guide. Society for Industrial Mathematics (1987)

4. Dongarra, J.J., Luszczek, P., Petitet, A.: The LINPACK Benchmark: Past, Present, and Future. Concurrency and Computation: Practice and Experience **15**(9), 803–820 (2003)
5. Petrini, F., Kerbyson, D.J., Pakin, S.: The case of the missing supercomputer performance: Achieving optimal performance on the 8,192 processors of ASCI Q. In: Proceedings of the 2003 ACM/IEEE conference on Supercomputing, p. 55. IEEE Computer Society, Washington, DC, USA (2003)
6. Snell, A., Goldfarb, D., Willard, C.G.: Designed to scale: The Cray XT5 family of supercomputers. White paper (2007)

Prerequisites for the Productive Usage of Hybrid Systems

Danny Sternkopf and Martin Galle

Abstract The employment of a hybrid system can be very advantageous to increase the efficiency of coupled applications. This paper addresses the evaluation of the HLRS environment as a hybrid system. HLRS operates different architectures like NEC SX-8 (Vector), NEC TX-7 IA64 (SMP) and PC cluster based on Intel EM64T. The different and independent computing systems of this environment can be coupled and used as a hybrid system, if a respective set of software components is provided. The evaluation presented in this paper focuses on Batch systems, Global File systems and MPI/coupling software.

1 Introduction

Different HPC applications obey their own physical laws. A coupling of different applications aiming at a multi-physical simulation has to combine the different physical laws.

A prominent example for application coupling is the fluid-structure-interaction, see also Fig. 1, where a fluid interacts with a solid structure, exerting pressure that may cause deformation in the structure, which in turn has an impact on the flow-field. For example: The air around an airplane wing causes the wing to deform which changes the air pattern around it.

Beside of this there are also different characteristics for:

- Simulation and re-meshing
- Ocean and atmosphere
- Aero-acoustics (Noise generation and propagation)
- Flow and chemistry [10]

Danny Sternkopf · Martin Galle
NEC Deutschland GmbH, Hessbrühlstrasse 21b, 70565 Stuttgart, Germany, e-mail: {dsternkopf, mgalle}@hpce.nec.com

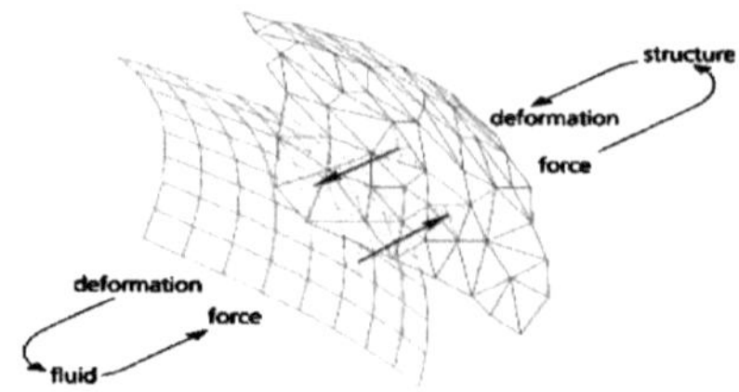

Fig. 1 Fluid-Structure Interaction

Coupling of different HPC applications is a natural approach for dealing with the simulation of very complex systems. It allows to describe the whole complex as one entity. These coupled applications consist of two or more different programs which run either simultaneously or alternating. The advantages to run coupled applications on heterogeneous platform are:

- Codes are well suited for specific architecture like vector or scalar due to their characteristics.
- Some ISV codes are not available on every platform.

For example the Distributed European Infrastructure for Supercomputing Applications (DEISA) project supports and provides coupled applications. Six coupled applications have been deployed so far [1].

2 HLRS System Overview

The evaluation system comprises 72 vector nodes SX-8, two TX-7 SMP nodes and a cluster of 210 Intel Xeon nodes as shown in Fig. 2.

1. SX-8 (OS: Super-UX 15.1) The SX-8 is a multi-node vector system where each node has 8 CPUs and 128 GB of main memory. The nodes are connected to a high performance network switch called IXS which allows a very good scalability. This system is the most important part of the HLRS installation and provides the highest compute power. Various simulation programs run on it.
2. TX-7 ia64 (OS: SLES 9 SP3) The TX-7 are the front end nodes for the SX-8. One node consists of 32 Itanium2 CPUs and provides up to 512 GB of main memory. These machines are based on the NUMA architecture and they do not have any special network because they are single SMP systems. They are used as compute systems for jobs requiring a huge amount of main memory as pre-processing (mesh generation, discretization) and post-processing (visualization) applications.
3. PC cluster (OS: Scientific Linux 4.1) This cluster consists of 210 nodes carrying two Intel XEON CPUs (em64t) and 2/8 GB of main memory. They are connected

with a Voltaire Infiniband high-speed network. Applications which suffer from poor vectorizability run on this cluster.

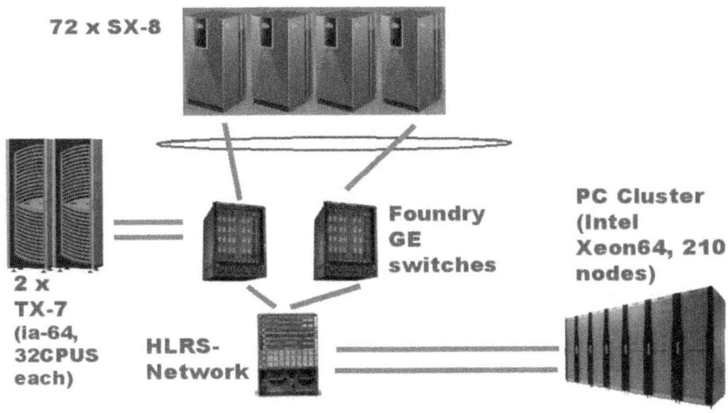

Fig. 2 HLRS System Overview

From an application point of view, the system setup must provide different levels to ease the usage of the complete system.

- Level 1: Shared File System
  All machines have (fast) access to one single file system.
- Level 2: Message Passing
  MPI which provides interoperability features like little to big endian data conversion.
- Level 3: Shared Memory
  Mechanism providing a fast direct access to remote main memory (like RDMA).

The current evaluation focuses on Level 1 and Level 2, but not on Level 3.

3 SW Components for a Hybrid System

To achieve these levels the following software components have been implemented and evaluated on the HLRS system:

- Batch system
  It is a requirement for the productive usage of the hybrid system. The allocation of resources must be scheduled centrally and dedicated.
- Global file system
  This is a requirement for Level 1. NEC GFS is already running on two out of the three different systems. Therefore it is chosen to integrate the PC Cluster into it as well.

- MPI
  This is a requirement for Level 2. PACX-MPI from HLRS is already available on all platforms. As an alternative MpCCI is also tested during the evaluation.

3.1 Batch System

There is NQSII together with two schedulers, Jobmanipulator and NQSII scheduler, within one batch server running on the SX-8 and TX-7. On the PC cluster the Torque (OpenPBS based) batch system with Maui scheduler is running.

Both TX-7 nodes are configured in shared batch mode. Also a few SX-8 nodes are configured in shared batch mode, while the others are used in dedicated mode. On the PC cluster nodes can be allocated dedicated only. Both batch systems, NQSII and Torque, are running on dedicated batch servers.

There are several possible solutions to have a central scheduling mechanism. Some are listed below:

- Meta scheduler [5]
  This is just a scheduler like Gridway [2] or Moab [4] which combines multiple distributed resource managers. There is no scheduler which supports NQSII. It would require additional effort to get NQSII integrated into such kind of meta schedulers.
- NQSII or Torque
  They are already in use on part of the system. There is no NQSII available for the Linux version Scientific Linux 4.1 running on the PC cluster. Beside of this additional licenses would be needed for the NQSII batch server to run 210 more clients. The Torque batch system [4] would have to be ported to the SX-8. This might be feasible, but it is not clear how much resources would be required.
- Third-Party software:

  - PBSPro [9]
    It is available on all platforms. A license is needed.
  - LSF [3]
    It is available for the Linux versions but not for Super-UX. A license is needed.
  - N1GE(SGE) [7]
    It is available for Linux versions. But it must be ported for Super-UX. No license is required.

PBSPro is chosen for evaluation because it is available and supported for all relevant platforms including the SX-8. NEC maintains a close contact to Altair to get support if needed. PBSPro does not require much effort to get it started, porting is not needed and configuration is straight forward. PBSPro brings along a batch server and a batch scheduler [9].

For this evaluation PBSPro version 9.0 is installed and configured on two PC cluster nodes, two SX-8 nodes and one TX-7 machine. With this test environment it was possible to schedule jobs using all three different systems. The PBSPro GUI xbsmon used for monitoring the batch system is shown in Fig. 3.

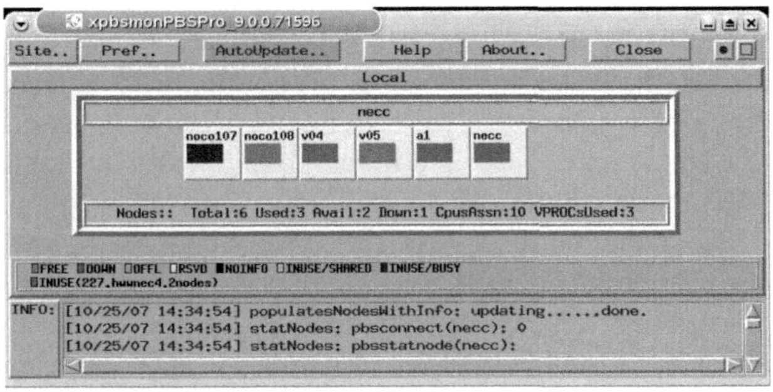

Fig. 3 PBSPro GUI: xpbsmon

During evaluation, the other batch clients were disabled on these nodes. Only PBSPro was active at that time to avoid multiple allocations of the same resources.

3.2 Global Filesystem

NEC GFS runs on the SX-8 and TX-7 which have access to several single coherent file systems. These are one 80 TB scratch file system, two 9 TB HOME file systems and two file systems for special purposes. The PC cluster is not integrated and does not have access to them.

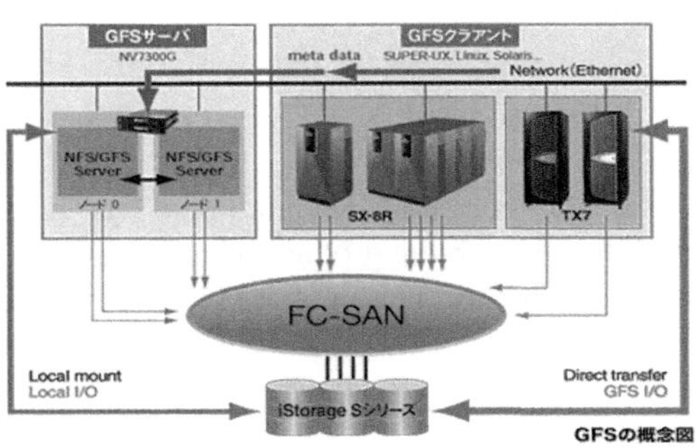

Fig. 4 NEC GFS Overview

NEC GFS (gStorageFS) is a high performance network file system working as client-server concept, see also Fig. 4. Data is directly transferred between client servers and storage devices. Only meta data traffic goes through the file server which is the NEC NV7300G (IA-32 based NAS head). It is a SAN-based (Storage Area Network) file system that takes advantage of a Fibre Channel infrastructure.

The idea was to port the GFS client to Scientific Linux 4.1 on the PC Cluster nodes in order to integrate the cluster into the GFS environment. The GFS development in NEC Tokyo provided a binary package and the source code for the SLES9 (x86_64) version of GFS. This GFS client called mGFS should use the Infiniband network installed on the PC cluster to access the existing FC SAN network with the help of a FC Router for data traffic, see also Fig. 5. This FC Router is an I/O module with four 2 Gbps FC ports installed in the Voltaire ISR9288 Infiniband switch. The first step was the integration of the FC Router into the existing FC SAN network. The FC SAN fabric consists of several Cisco MDS 9000 switches which are configured in two hierarchical levels.

Another complication was generated by the fact that the PC cluster and the rest of the environment are installed in different buildings. There are about 700 meters patched connections between both buildings.

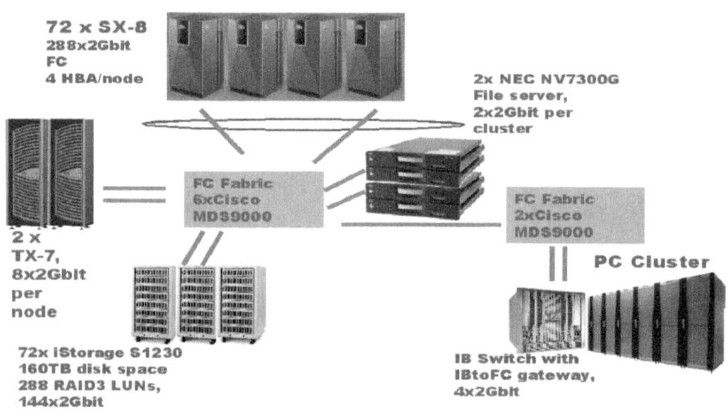

Fig. 5 GFS FC SAN configuration

The next step was to run and test the SLES9 mGFS client on an Opteron server which was integrated into the PC cluster. This was done successfully without any problems. The following configuration shown in Fig. 6 was used at the beginning of the evaluation which didn't disturb the production environment.

The Azusa (IA64 SMP platform) was used as the file server. This node serves the four SX-6 GFS file systems. One of them was out of production and could be used dedicated for testing.

Then the mGFS port to Scientific Linux 4.1 started which mostly needed kernel adoption for the 2.6.9 kernel version. The original code was developed for SLES9 and SLES10 kernel versions which are based on other releases. A data verification test to validate the functionality of mGFS has been carried out successfully after an IB HCA FW update was done. 288 MB/s for read and 136 MB/s for write was measured on one client node which is quite similar to the results on the Opteron

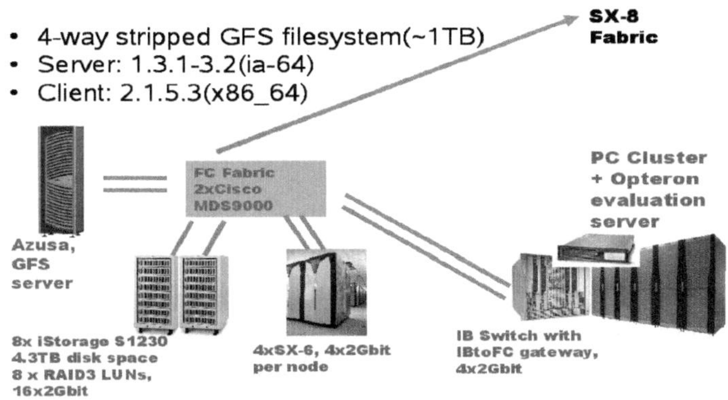

Fig. 6 GFS evaluation environment

server under SLES9. Next steps were to move to GFS production environment, see also Fig. 7, and to perform stability and scalability tests with a higher number of PC cluster nodes.

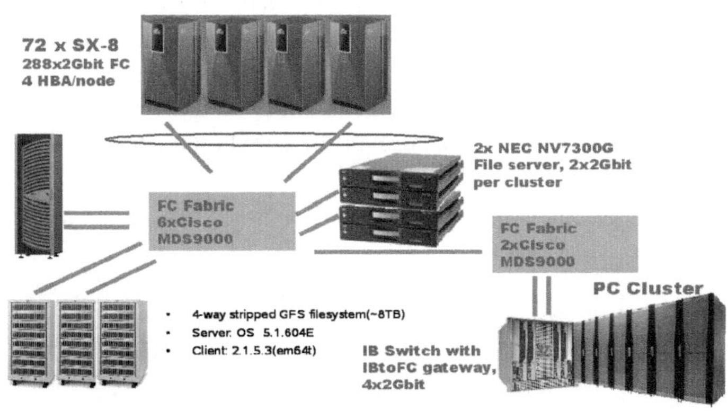

Fig. 7 GFS production environment

The NV7300G is used as file server now. Three clients, two PC cluster nodes and the Opteron server, were configured for concurrent usage. The tests revealed several problems with limitations of the FC Router and with problems with the iSCSI initiator software which could not be solved. Finally Voltaire offered to test the SR4G which is next generation of the FC Router. The Storage Router (SR4G) has four 4 Gbps FC ports and two Infiniband ports and is a own box (server). So the FC Router was replaced by the SR4G and those problems were solved. The stability and scalability were noticeable better. The tests were done on up to 170 PC cluster nodes so far. A single node achieved 386 MB/s read and 220 MB/s write. On 8 nodes 707 MB/s read and 456 MB/s write are measured. The intention is to use that mGFS client on the PC cluster in production. So there will be a smaller GFS file system which is used on the PC cluster and later on the SX-8 as well to ease the run of coupled applications and the workflow in general. Further plans are to integrate the PC cluster completely into the current productive GFS environment.

3.3 MPI

HLRS has developed a machine independent MPI extension called PACX-MPI [8]. It can be used to run MPI applications over different architectures. It uses the native MPI on every system for internal communication and a special protocol for external communication between different systems to take care about interoperability like data conversion. There are already some applications for which SX and PC cluster were coupled with PACX-MPI successfully [11].

Another approach is to use MpCCI. This is the Mesh-based parallel Code Coupling Interface (MpCCI) which is developed by Fraunhofer Institute Algorithms and Scientific Computing (SCAI). MpCCI carries out a data interpolation between meshes belonging to different simulation codes using a socket-based server–client communication. MpCCI acts as server and the API of the simulation code is the client. A pair of server and client for every simulation code has to be started. The internal communication between the servers uses MPI. Information exchange between the server and client is done by remote shell command like RSH/SSH. MpCCI does not interfere with internal parallelization of the simulation code based on MPI [6].

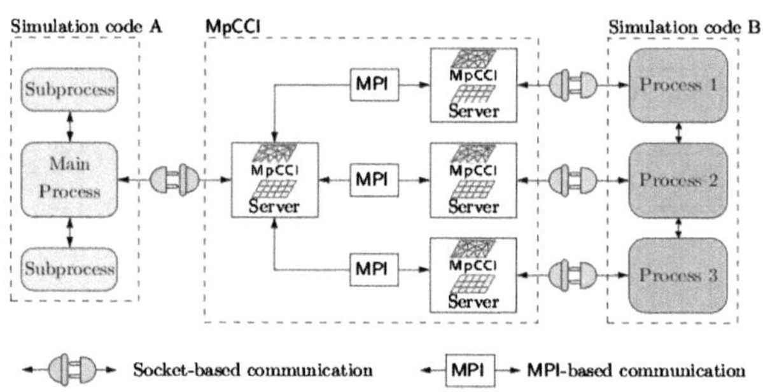

Fig. 8 MpCCI communication

This method was used to couple two applications, FENFLOSS (Flow simulation) running on the SX-8 and Abaqus (structure analysis) running on the PC cluster [12]. The FENFLOSS code was adapted to the SX-8 architecture within another TERAFLOP Workbench activity. Abaqus is a commercial ISV software which is available only for PC Cluster and not for the NEC SX-8. Therefore the MpCCI code adapter was ported on the SX-8 which is used by the API. The API is written in C and can be used in C, C++ and Fortran codes. The user could successfully run this coupled application with MpCCI [13].

4 Conclusion

The HLRS system can be regarded as a hybrid system. The required software components are being made available. There are at least two applications which run in hybrid mode so far.

References

1. DEISA — Distributed European Infrastructure for Supercomputing Applications. http://www.deisa.eu
2. GridWay. http://www.gridway.org
3. LSF. http://www.platform.com
4. Maui/Moab and Torque. http://www.clusterresources.com
5. Meta Scheduling. http://en.wikipedia.org/wiki/Meta-scheduling
6. MpCCI. MpCCI 3.0.6-12 Documentation, PDF version August 30 (2007) http://www.mpcci.de
7. N1GE. http://www.sun.com/software/gridware
8. PACX-MPI. http://www.hlrs.de/organization/amt/projects/pacx-mpi
9. PBSPro. http://www.altair.com/pbspro
10. Galle, M., Bönisch, T., Benkert, K., Borowski, S., Haberhauer, S., Lammers, P., Svensson, F., Tiyyagura, S.R., Resch, M., Bez W.: The HLRS–NEC Teraflop Workbench — Strategies, result and future. In: M. Resch, T. Bönisch, S.R. Tiyyagura, T. Furui, Y. Seo, W. Bez (eds.) High Performance Computing on Vector Systems 2006, Springer (2006)
11. Klimach, H.: Coupled Simulation of aero-acoustic applications. 7th Teraflop Workshop November 2007, Sendai, Japan (2007)
12. Lippold, F.: Fluid-structure interaction using MpCCI 3.0.5 with ABAQUS and an in-house CFD-code on distributed resources. In: Proceedings of the 8th MpCCI User Forum, St. Augustin, Germany (2007)
13. Lippold, F.: Coupled applications on multiple HPC-Platforms. NUG-XX, Wien, Austria (2008)

II
Multiscale and Multiphysics Simulations

Complex Numerical Simulations in Production Techniques

A. Scheibe and E. Westkämper

Abstract The modelling of manufacturing processes is approached in the professional communities as one of the essential, long-term demand, which requires the collaboration between large spectrums of research areas. This paper presents our innovative approach for a comprehensive and holistic factory simulation model. "Multi-scale" in this context not only signifies the spatial and temporal scales within separate manufacturing processes, but also the different scales of all running processes in the whole manufacturing enterprise, called factory, as well as the different scales in the model itself. Therefore, for the purposes of the virtual and dynamic representation of the factory as a whole, at different scales or levels of abstraction, several heterogeneous modelling approaches have to be coupled with each other. In order to illustrate the usability of this methodology we gave at the end of the paper a complex scenario in the field of the automotive coating industry.

1 Problem Statement and Motivation

The manufacturing enterprises are confronted today with new models of competition and new modes of operation. They have to provide competitive industrial goods and support services at decreased prices with high quality and overcome the customers' expectations. In order to remain competitive or to survive, they have to adapt permanently to the needs and requirements of markets and economic efficiency. In pursuit of this objective, new and innovative methods of development and permanent optimisation of the manufacturing systems have to be constantly established [5]. This requires the development and mastering of highly efficient simulation algorithms and tools which support all areas of manufacturing engineering [6].

A. Scheibe · E. Westkämper
Institute of Industrial Manufacturing and Management (IFF), Nobelstr. 12, 70569 Stuttgart, Germany, e-mail: {andreas.scheibe, westkaemper}@iff.uni-stuttgart.de

Today, there is a lack of formal methods to model and simulate manufacturing systems. The simulation is limited to geometry in native formats, or to models made for specific applications. There is a need for reusable models and specifications that can be used for more than one application. Another identified gap in the simulation domain is the definition of new formalisms and modular modelling for distributed and multi-domain simulation of manufacturing systems in order to have integrable and reconfigurable simulation models used to optimize the complex manufacturing systems.

Motivated by these challenges, we set up our main research goal, the real-time simulation of integrated manufacturing systems, which are characterised by a multitude of heterogeneous parameters and autonomous units. Manufacturing networks and factory structures as well as the behaviour of manufacturing systems and mechatronic systems such as robots, machine tools and manufacturing factories have to be to be examined throughout their whole life cycle [1, 7]. The approach presented in this work supports the manufacturing enterprises to face these challenges by proposing new and innovative modelling and simulation methodology. This paper presents theses concepts and the first results.

2 Multi-Scale Modeling and Simulation of Manufacturing Systems

2.1 Vision and Scientific Impact

Modelling and simulation of complex networks of manufacturing systems is regarded as one of the central challenges for research in manufacturing technology. Adaptable simulation networks of cooperating and semi-autonomous software systems will lead to a fast market launch for new products as well as to shortened set-up times of production systems. Our work in the special research area of production engineering and manufacturing technologies aims in the next few years towards a resolute parallelisation of the development and set-up processes taking place on an interdisciplinary simulation network platform. With real-time systems, a conclusive transition or rather a practical link from the virtual to the real world will be developed. Controllable, new manufacturing processes will result from a combination of different production processes by means of hybrid and multi-scale simulation technologies. The computer-aided modelling, optimisation and prediction of complex phenomena in physical processes and factory systems will adopt a key position in the future, especially in the area of production engineering and manufacturing technologies.

2.2 Challenges, Theoretical Considerations and Definitions

The particular challenge of modelling and simulating manufacturing systems consists not only in determining the interactions of the different parameters and influencing factors but also in representing them in real-time simulation models. The term "real time" relates in this context to a fast reaction to arising events as well as the time-deterministic calculation of plant behaviour for the control-coupled simulation.

The multi-scale simulation of manufacturing systems aims to connect different computational and data model based simulation approaches through a continuous information flow, higher order equation systems and universal interfaces. The goal of this connection is the bridging of the discrete and numerical simulation, represented as the diagonal in Fig. 1, and of the areas of application in the manufacturing domain, as well.

In our approach, under the notion "multi-scale" not only the spatial and temporal scales within separate manufacturing processes is understood, but also the different scales of all running processes in the whole manufacturing enterprise, called factory, as well as the different scales in the model itself. Therefore, for the virtual representation of the factory as whole [7], several heterogeneous modelling approaches have to be coupled with each other at different levels of abstraction.

In order to model and simulate comprehensively at all scales presented in Fig. 1, the following challenge has to be overcome: if one regards today's' conditions of the simulation of technical processes and factory-organizational operational sequences, then this can be based on completely different, decoupled data and computational models. There is no constant information flow between these scales. Therefore, the main goal is to solve the integration between the heterogeneous simulation models,

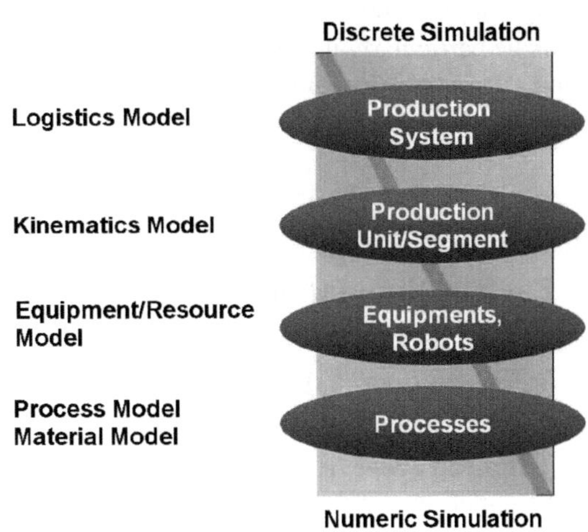

Fig. 1 Connection of different computational and data model-based simulation approaches

as represented by the diagonal in Fig. 1. This heterogeneity consists in the migration of simulation models from numeric simulations for the process and material modelling towards the discrete simulation of logistics. The complexity arises when modelling the horizontal scales of the factory as whole, beginning with the technical manufacturing processes over equipments, robots, production systems, segments to a network of production systems. As a conclusion, the following aspects and scales have to be regarded:

- all levels of factory structures (manufacturing processes, machines, manufacturing system and network of manufacturing),
- several areas and concepts related to mathematics, physics, chemistry, engineering, economic science,
- several simulation methods (molecular dynamics, finite elements, event-oriented simulation),
- spatial expansion (atomically, microscopically, mesoscopically, macroscopically) and
- temporal expansion (nanoseconds, seconds, hours, days, weeks, years).

All processes considered here are based on different numeric computations and run offs on different time scales. To interconnect them and thus to draw a bow of the technical processes up to the complete production system stretch, is the central setting of tasks for the next years. Figure 3 shows exemplarily the different temporal scalings of the simulation on different levels of the factory structure. While the physical procedures during the technical processes usually take place within a range from milliseconds up to few minutes, the factory organizational operational sequence in the entire production system extends up to several weeks and years. Figure 2 displays knowledge-based models, the kinetical and logistical simulation [2] for the different levels of the factory structure and/or the operational areas as well as the different numeric solutions methods like the molecular dynamics (MD), Monte Carlo (MC) and finite element method (FEM),

The application areas of multi-scale simulation for modern manufacturing systems lie particularly in: a) factory and logistics planning, b) work and process planning, c) construction of the operational funds, d) programming and control, e) processing and f) quality management.

3 A New Concept Towards a Comprehensive and Holistic Factory Simulation Model

3.1 Theoretical Aspects

As described above, it is the aim of our research activities to couple the modelling and simulation of technical and logistic processes of manufacturing systems, in particular in the area of surface technology. Discrete and continuous simulation approaches, differing one from another as a matter of principle, have to be combined and referred to.

Only the coupling of the process models (at the lowest level) with the time-discrete and event-oriented model (at a highest level) via several scaling levels (conf. Fig. 1) will allow for a holistic and integrated approach and the identification of interdependencies and effects within the manufacturing system. To implement this coupling, several different approaches will be investigated and enhanced.

1. At the lowest level of material and process simulation shown in Fig. 1, physical, chemical and atomistic models based on similar differential equations are integrated into a comprehensive computational model and solved as one, and if applicable, in parallel. The general transportation equation, for example, allows the description of processes in fluid mechanics, electrodynamics and thermal conduction as well as of mass transport. The single coefficients of the general transportation equation can be set in the code individually and dynamically for the respective problem. For solving complex problems, several differential equations are integrated into a single system with several interdependent variables. In addition, the use of a common computational network simplifies process chain simulations over several steps, such as the simulation of applying paint on a car body and subsequently drying it in an industrial furnace.
2. Process modules which may not be coupled by a common approach will be combined through common input and output variables at defined interfaces. Molecular-dynamic simulations, for example, are usually based on stochastic computation models, and not on differential equations as is the case with most deterministic process models. However, both simulation approaches make use of common process parameters, i.e. input and output variables. These are used in interfaces and as constraints/boundary conditions, respectively.
3. At the level of kinematical models, multi-body systems are described by complex differential equations for movements and the so-called inverse kinematics. For

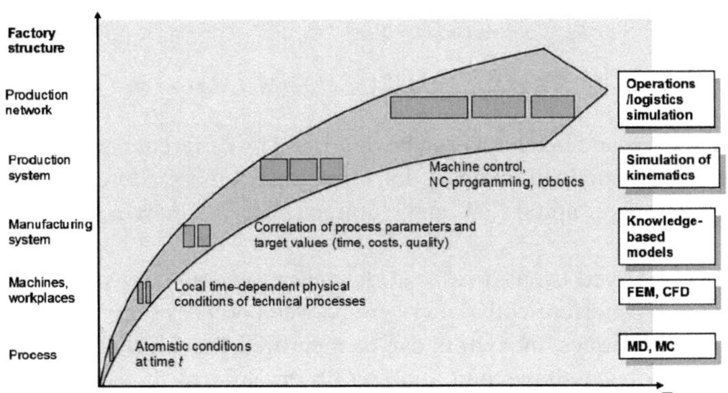

Fig. 2 Different scales in time, of factory structures and models

coupling motions with heat flows, forces and moments, extended FEM and CFD programmes are used.

4. The correlation between quality features, process parameters and their interactions with the manufacturing system is being described using knowledge based systems, intelligent and multi-agent application systems, if no direct physical or economic description is appropriate.
5. Another coupled approach at the highest level of the manufacturing system is a vectorial description of all necessary tolerances and logistical characteristics in a multi-dimensional space. Within this space, various subspaces and areas may be defined describing the given state of the manufacturing system and allowing the evaluation of both technical processes and logistic chains. This approach will be illustrated in Sect. 3.2.

3.2 Vectorial Modelling and Simulation Approach Based on Business/Organizational Parameters and Technical Tolerances

Our vectorial approach is represented through the following theoretical and technical considerations. The state of the entire manufacturing system in a specified moment in time, $MS(t)$, can be described by the corresponding production system $PS(t)$, the logistics system $LS(t)$ and time as

$$MS(t) = f(PS(t); LS(t)) . \tag{1}$$

The production system

$$PS(t) = f(M1(t), M2(t), \ldots, Mn(t); P1(t), P2(t), \ldots, Pn(t)) \tag{2}$$

specifies the involved machines $M(t)$ and manufacturing processes $P(t)$, respectively. The logistics system models order sequence $OS(t)$ per process and the respective stocks $sto(t)$ as

$$LS(t) = f(OS1(t), OS2(t), \ldots, OSn(t); sto1(t), sto2(t), \ldots, sto-n(t)) . \tag{3}$$

The individual processes and machines can be described by different parameters. The order sequence depends on lead time and delivering faithfulness. Store size can be described against lead time, capital costs and required space. All these parameters are time variable.

First, the output is observed event-driven, at discrete moments after each process. Thus, it is possible to interfere with the system between process steps. Second, with in situ measuring techniques, the output can be measured and adjusted already during the process which necessitates a continuing observation of the production system.

With this holistic model approach the interaction between technical processes and logistics can be detected and considered. Technical tolerances as well as logistic parameters span a tolerance and parameter channel over the process sequence (Figure 3). All output data (such as dimension, form, location, waviness, roughness, working load, capacity, lead time and delivering faithfulness of each process step) should lie within this channel, in order to define the appropriate input (e.g. process parameters, order sequences for the following process).

The production system aims at fulfilling the stipulated figures/values at least by the last process step (products). If, on the way, one process step diverges from these values, appropriate measures can be taken and simulated using the holistic model approach, such as adjusted process parameters or changed order sequences.

These new general conditions can also have impact on the real process and rule back into the parameter channel. If, for example, the minimum coating layer is not reached by the first strike it can be adjusted by the following second application by means of an adjusted process conduct under the new general conditions.

The actual coating process, however, has such complex impact correlations and many different parameters, that adjustment and optimization can only be realized with the help of a physical (layer-based) coating simulation.

With this innovative procedure, new products and variants (e.g. effects of process ability of production processes on lead time, re-operation rates and resulting piece costs) can be modelled and optimized even in the run-up phase.

In this process, commercial software tools like eMPlant, Witness or Quest play a major role for material flow simulation and logistics simulation. With these tools, parameters such as lead time, working load of machines, plants, vehicles and staff, can be analyzed at present based on statistical and estimated data. However, this happens without consideration of technical tolerances. Also, the logistics tool gives no feedback to the machine and the technical process. These deficits are to be solved with the presented approach.

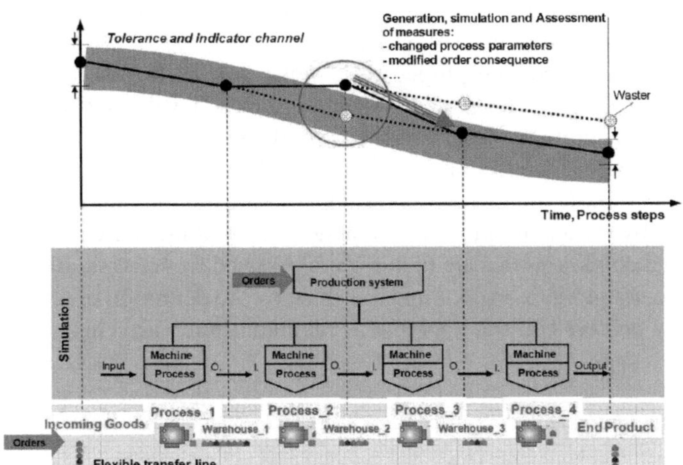

Fig. 3 Illustration of coupling technical tolerances and logistical parameters

4 Complex Example in Automotive Coating Industry

The modelling and simulation of the relevant processes in coating technology and the integration of several heterogeneous models into the "Virtual Factory" [4] are based on the principles of physics (computed with Finite Elements (FEM), Finite Volume (FVM), Molecular Dynamics (MD) and Monte Carlo (MC)) as well as different logistical and material flow simulations. The latter have to be drawn on and linked together across several scaling levels. Table 1, which has to be read bottom-up, presents different simulation approaches, structured according to the factory scales, some relevant features and areas of application. The special objectives in this task lie in the efficient coupling of various independent mathematical and physical differential equations and of numerical methods of resolution. For the simulation of the spray jet, the flow conditions described by the Navier-Stokes equations with conservation of energy, mass and momentum as well as parts of Maxwell's equations for the description of electrical field and charge transport have to be solved. The spraying of individual paint droplets is subsequently calculated using the differential equations of motion with a Lagrange particle tracking method taking into consideration special turbulence models and simultaneously solving the space and time-dependent differential equation for the electrical charging [3, 8] of each individual paint droplet (levels 1 and 2). The simultaneous solving of the differential equations can only be done numerically by a coupled iterative calculation method, like for instance the Finite Volume Method.

Only through coupling different approaches by an iterative solver the interaction of flow expulsion caused by the droplets and the induction of space charge and hence of the electrical field can be described with sufficient accuracy. In the extended task posed here for an unsteady resolution of the problem taking into consideration evaporation, drying (level 3) as well as film formation, further differential equations from thermodynamics and materials science as well as terms for the description of time-dependent material behaviour have to be added.

The prediction of quality and film properties is done on level 0 by a separate molecular dynamics simulation, where subsets of the results of level 2 and 1 serve as initial conditions. The coupling of level 0 happens at a defined process point of the vectorial approach of mapping the technical tolerances and company organisational parameters (see Fig. 3). On level 4, the processing of robot motion and machine behaviour takes place by means of "inverse kinematics" in combination with FEM simulations. The initial conditions that are required are obtained from the simulation results on level 2. The coupling to levels 4 and 5 is done by the definition of technical parameters and by process control, applying a continuous tolerance channel, as shown in Fig. 3.

Table 1 Overview of different simulation models on different levels of scaling and abstraction

Simulation Level & Example	Target Values	Initial Conditions	Single Results	Simulation Models
Level 5: Production systems 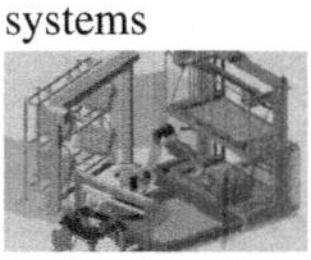	Factory planning, interaction with other manufacturing systems	Conditions of manufacturing systems	Overall result: Virtual Factory	Event-driven models, Virtual Reality, tolerance channel
Level 4: Manufacturing systems 	Machine and plant layout	Process parameters, plant data and characteristics	Material flow, logistics, capacities	Event-driven models, models of material flow and kinematics (robotics), ergonomics, tolerance channel
Level 3: Drying (macroscopic) 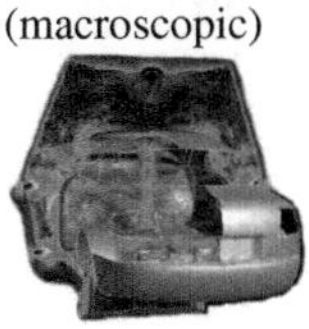	Degree of drying, hardness, surface structure (e.g., orange peel), distortion of parts, optimisation of process control in drying	CAD geometry, process parameters, paint film thickness, material parameters	Heating curves, heat distribution and radiation, paint film quality, optimised process control, energy costs, oven dimensions	Equations for material transport by thermodynamics and convective flow (CFD, FEM), mechanical strength (FEM), neural networks, knowledge-based systems, tolerance channel
Level 2: Film formation (macroscopic) 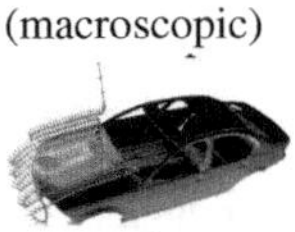	Total film thickness, paint consumption, optimisation of process control using different application techniques, global quality criteria	CAD geometry, process parameters, film properties and quality criteria	Local mass flow, paint charging (flow of current), momentum of individual paint droplets, energy costs, dimensions of coating line	Transport equation for particle-laden flows (CVD), Maxwell equations for electrodynamics (FEM), thermodynamics, neural networks, knowledge-based systems
Level 1: Paint film levelling (mesoscopic) 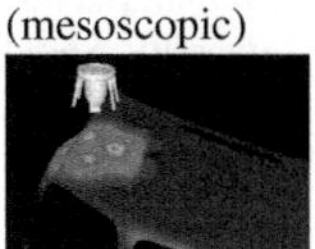	Film thickness distribution, paint film levelling, evaporation of solvents, film formation, film properties, quality criteria	CAD geometry, material properties and parameters (rheology), initial paint film surface structure	Optical and mechanical quality criteria, e.g., film thickness, gloss, adhesion	Transport equations, free surface flow (CFD), Maxwell equations for electrodynamics, thermodynamics, material simulation (paint chemistry)
Level 0: Droplet impingement (microscopic) 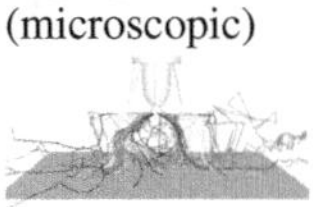	Film properties	Droplet impingement, momentum of paint droplets, material properties	Atomistic initial conditions for optical and mechanical quality criteria	Statistical approaches MD, MC

5 Conclusions and Future Work

The multi-scale simulation of a factory represents a great challenge which manufacturing systems face in order to be flexible and permanently adaptable to the new requirements of markets and economic efficiency. This paper presented our approach called comprehensive and holistic multi-scale simulation. It takes into consideration complex aspects of manufacturing processes like spatial and temporal aspects as well as the scales of running processes in the whole manufacturing enterprise. The innovative methodology proposed couples different physical-based simulations with logistic and flow of material simulations. The presented vectorial modelling and simulation concepts based on business/organisational parameters and technical tolerances represents a valuable tool that will be used for modelling and simulating the manufacturing systems, especially in the automotive coating industry. An overview of different simulation approaches at several scales has been presented in detail. Currently our work focuses on the levels 1, 2 and 3. In the future, we intent to explore the uncovered levels 0, 4 and 5 in an interdisciplinary cooperation with other recognised research units and institutions of the Universität Stuttgart.

References

1. Aldinger, L., Constantinescu, C., Hummel, V., Kreuzhage, R., Westkämper, E.: New approaches for “advanced Manufacturing Engineering”. Scientific management, factory life cycle, factory as a product and advanced Industrial Engineering (aIE). In: wt Werkstattstechnik online, Springer-VDI-Verlag wt-online **3** (2006)
2. Mitschang, B., Westkämper, E., Constantinescu, C., Heinkel, U., Löffler, B., Rantzau, R., Winkler, R.: A flexible integration of layout planning and logistics simulation through data change propagation. In: Proceedings of the 36th CIRP International Seminar on Manufacturing Systems (CIRP ISMS 2003), Saarland University, Saarbrücken (2003)
3. Scheibe, A., Domnick, J., Ye, Q.: Simulation of the behaviour of sprays from high-speed rotary bell atomizers with external charging. In: 17th Annual Conference of ILASS-Europe on Liquid Atomization and Spray Systems, Zürich (2001)
4. Scheibe, A.: Anwendung der Beschichtungssimulation in der Praxis. Simulation in der Lackiertechnik. In: Proceedings of Fraunhofer IPA/IFF-Workshop, Stuttgart (2004)
5. Westkämper, E., Winkler, R.: The use of system modelling for the intelligent planning, scheduling and control of agile manufacturing. In: J. Kim (ed.) 35th CIRP-ISMS manufacturing technology in the information age, 12.-15. May 2002, Seoul National University, Seoul (2002)
6. Westkämper, E., Pfeffer, M., Dürr, M.: Partizipative Fabrikplanung mit skalierbarem Modell, In: wt Werkstattstechnik **94**(3), 48–51 (2004)
7. Westkämper, E, Constantinescu, C., Hummel, V.: New paradigms in manufacturing engineering: Factory life cycle. In: Annals of the Academic Society for Production Engineering. Research and Development, XIII/1, vol. XIII(1) (2006)
8. Ye, Q., Scheibe, A., Domnick, J.: Numerical simulation of electrostatic spray-painting processes in the automotive industry. In: E. Krause: High Performance Computing in the Science and Engineering, ISBN 3-540-22943-4, pp.261–275. Springer, Berlin (2005)

Multi-scale and Multi-physics Applications — User Requirements for Future Applications

Sabine P. Roller

Abstract Multi-scale and multi-physics simulations have been upcoming for the past years in different application scenarios from various scientific areas. Typically they result in coupled applications where different modules are — tightly or loosely — coupled to solve a problem that is closer to reality than a single-scale or single-physics application. To run these coupled simulations, several requirements are imposed upon the system, ranging from availability of coupling tools as well as management features like co- or conditional scheduling. This holds especially true when trying to map the different modules of a simulation to different hardware architectures, i.e. to a hybrid hardware environment. The reason for doing that is that a one-size-fits-all approach with forcing every module to the same machine might restrict the possible simulations too much. This is particularly the case if one of the modules is available on a distinct machine only — typically commercial codes with restricted licenses and availability.

1 Introduction

Multi-scale and multi-physics simulations have been upcoming for the past years in different application scenarios from various scientific areas. Typically they result in coupled applications where different modules are — tightly or loosely — coupled to solve a problem that is closer to reality than a single-scale or single-physics application. To run these coupled simulations, several requirements are imposed upon the system, ranging from availability of coupling tools as well as management features like co- or conditional scheduling. This holds especially true when trying to map the different modules of a simulation to different hardware architectures, i.e. to a hybrid hardware environment.

Sabine Roller
High Performance Computing Center Stuttgart (HLRS), Nobelstr. 19, 70569 Stuttgart, Germany, e-mail: roller@hlrs.de

The reason for trying to map different modules to a non-homogeneous platform is that a one-size-fits-all approach with forcing every module to the same machine might restrict the possible simulations too much. This is particularly the case if one of the modules is available on a distinct machine only — typically commercial codes with restricted licenses and availability.

The paper deals with upcoming topics for future applications and their requirements toward future systems from a user's and operational perspective. The background of this work is based on the experiences obtained in the Teraflop Workbench project. The concept of the project is to bring together researchers, having application know-how as well as a strong physical background, and experts from numerical mathematics, computer science and/or computer engineering to enable leading edge computational science in areas such as

- Automotive engineering,
- Aerospace engineering,
- Civil Engineering,
- Combustion,
- Turbo Machinery,
- Molecular Dynamics,
- Astrophysics,
- Medical Surgery and
- Bioinformatics.

2 A Glance at Past, Current and Future Simulation Applications

When looking at the developments in simulation technologies of past, current, and future applications, we can summarize the following facts:

- In the past, emphasis was placed on the development of simulation tools. The models were simple and had a small number of grid points and degrees of freedom (DoFs), respectively. The applications consisted of individual application runs doing forward simulation, i.e. calculated for some given input data the behaviour of the system in the end or within some given time interval (stationary or unstationary).
- Looking at today's applications and ongoing developments, we see a simulation based development. The goal is not to make a simulation simply run, but to use simulations as a state-of-the-art tool in developing products and devices. Applications are still mainly single-physics, single-scale simulations, but also multi-scale and multi-physics applications are present which couple the single-physics/scale codes as modules into a new application. In most cases, the coupling is specific to the individual application and code, but tendencies to develop more generic coupling tools are observed.

Another on-going development is design optimization. It can also be viewed as a coupled simulation, since several instances of a single code are called and the results are evaluated with respect to a goal function.
- When talking about future plans, the tendency is to use simulation to steer complete physical processes. This includes ensemble and backward/inverse simulations, i.e. correct initial and boundary conditions are calculated to achieve this behavior. Applications are medical computations, environmental behavior or industrial production chains.

3 Future Application Perspective

The observation of a wide variety of applications clearly shows that the way to get closer to reality is no longer marked only by higher resolutions, but also by more accurate modeling, i.e. by including more physical phenomena into the simulation. This can occur

- within one code using more equations (e.g. for chemical reactions in fluid flow),
- by coupling independent modules (e.g. fluid-structure interaction) or
- by running different simulations sequentially as workflow (e.g. medical surgery, interactive visualization).

In all cases, one observes that users often treat modules as a black box. Since their expertise is usually limited to one area of the coupled application, modules are often used like libraries. The second important observation is that coupling often involves a combination of an in-house code plus a commercial software. This has two implications:

1. The source code is partly not available. Thus, the coupling has to be realized via user-provided coupling routines and/or input/output procedures.
2. The fact that commercial software is usually only available for standard PC clusters is even more problematic. If all parts of the simulation have to run on the same system, the commercial tool determines the architecture to be used. This is a severe restriction for high performance computing.

It is illegitimate to assume that all modules in coupled simulations are suited for the same architecture. One can devise several combinations of applications and hardware platforms. Therefore, it takes careful thinking about the requirements of each module and how to each module individually to the best suited hardware platform.

4 Examples

4.1 Dynamic Coating Simulation

An example for process steering simulations is the dynamic coating simulation of complete car bodies, investigated at the Institute of Industrial Manufacturing and Management, University of Stuttgart: it analyzes the full production chain from static spray patterns on different geometries over dynamic pathways for automatization to the control of entire coating lines of an entire car body [7]. The vision of digital manufacturing is to unify simulation models in vertical direction ranging from processes to equipments/robots, to production unit/segment and finally to production system.

4.2 Ocean Water Plant

Fluid-structure interaction is a typical example for multi-physics applications. It is used at the Institute of Fluid Mechanics and Hydraulic Machinery, University of Stuttgart, to design and optimize propeller blades for ocean water plants. A total set of four individual codes is applied: a fluid solver, a structural mechanics package, a commercial coupling tool as well as a visualization and steering environment for the collaborative working suite in the design process [5].

Figure 1 demonstrates the interaction of the software packages, while Fig. 2 depicts the interactions of the user with the different resources if not all modules are located on the same resource. It is obvious that support is necessary for such complex simulation workflows.

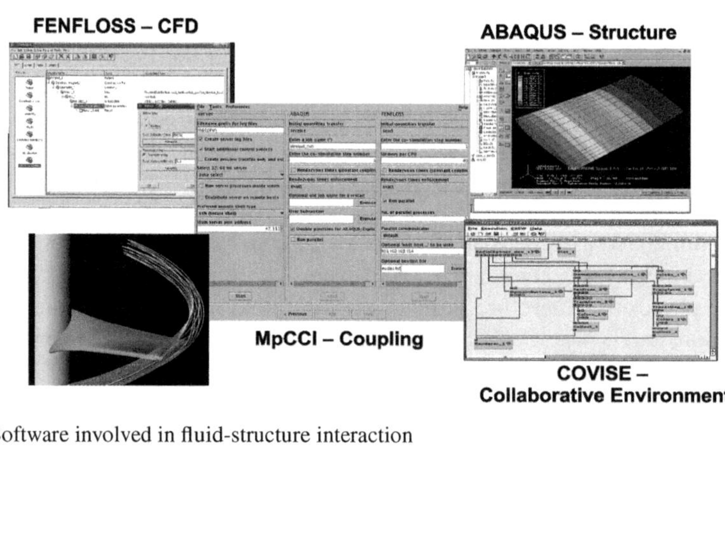

Fig. 1 Software involved in fluid-structure interaction

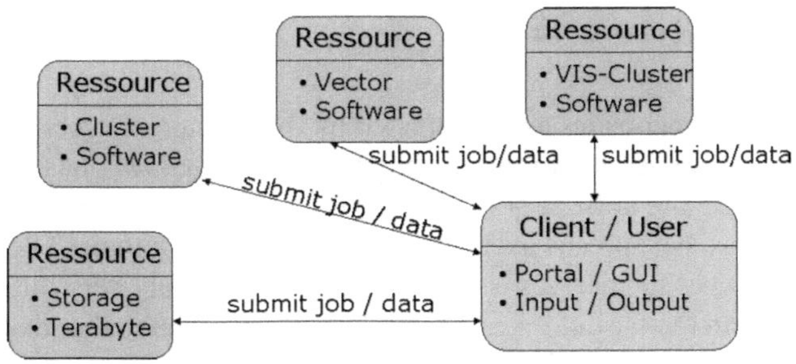

Fig. 2 Interaction between user and resource

4.3 Particle-loaden Flows

Particle-loaden flows are investigated in a joint research project by the Institutes of Space Systems and for Aerodynamics and Gasdynamics at the University of Stuttgart, the Institute for Pulsed Power and Microwave Technology at Research Center Karlsruhe, and the High Performance Computing Center Stuttgart (HLRS). The application is strongly coupled, but the different modules depicted in Fig. 3 introduce an additional hierarchy in the sense of instruction parallelism. Again, the requirements for the different modules might be competing [1, 2, 6].

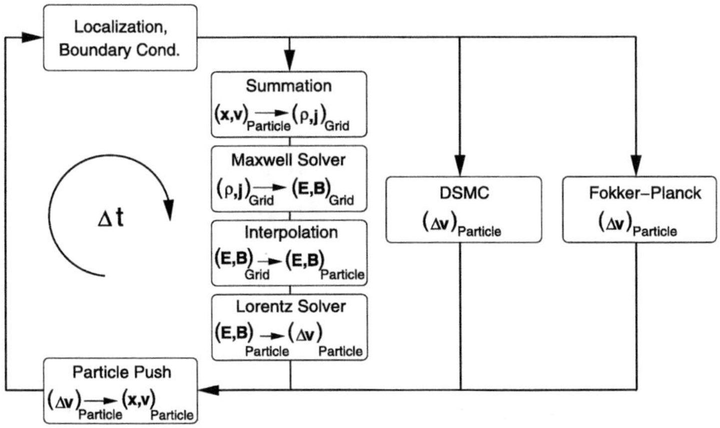

Fig. 3 Instruction-parallel cycle of the PIC+DSMC scheme

4.4 Multi-scale Aero-acoustics

Computational aero-acoustics as investigated by the Institute for Aerodynamics and Gasdynamics, University of Stuttgart, are typical multi-scale applications. The flow which is generating the noise is slow and carries a large amount of energy, i.e. large pressure variations, but spreads over a comparatively small domain. The propagation of the acoustic waves themselves is much faster, carrying only very little energy, but one is interested in the far field, i.e. a much larger computational domain.

A heterogeneous domain decomposition approach (see Fig. 4) is taken, where heterogeneity concerns the grid type (structured/unstructured), the coarseness of the mesh and thus the size of the time steps, but also the discretization methods (Finite Difference, Finite Volumes, Discontinous Galerkin), and even the equations (non-linear Navier-Stokes, non-linear Euler, linearized Euler) [4].

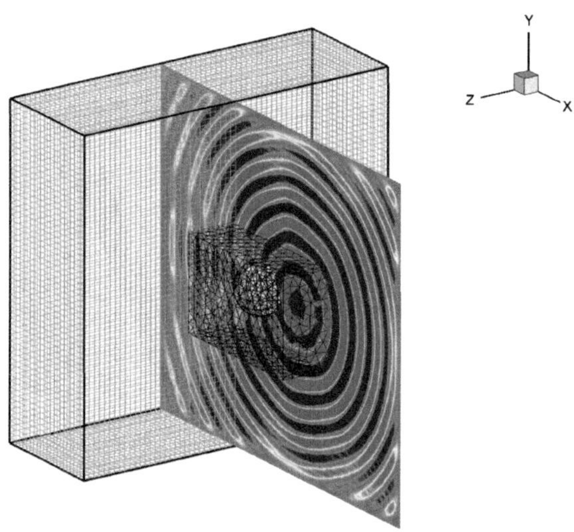

Fig. 4 3D computational aeroacoustics using heterogeneous domain decomposition

Typically, such a simulation consists of

- Innermost domain: unstructured fine grid, small time steps, non-linear equations, Navier-Stokes equations
- Middle domain: structured grid, coarser grid, larger time steps, non-linear equations, Euler equations
- Far field domain: structured, grid, coarsest grid, largest time steps, linearized equations, LEE or Lighthill analogy
- Coupling: mapping of meshes, interpolation, steering of sub-cycling processes

In this setting, the structured solver is well vectorizable and well suited for a vector architecture like the NEC-SX8. The unstructured solver is element-based and highly local, thereby less vectorizable. Thus, the best suited architecture for each task should be used.

The modules use a distributed memory parallelization with MPI as it is typical for domain decomposition methods. From the perspective of software development, this parallelization should be preserved even when distributing the individual MPI processes across distinct machines. HLRS developed the PACX library ([3]) — a tool to couple not the applications, but the individual MPI libraries on distinct machines. Its mode of operation is depicted in Fig. 5.

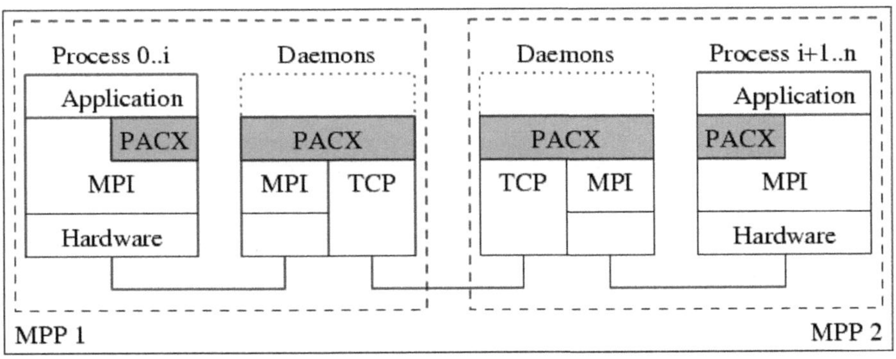

Fig. 5 Coupling of MPI libraries on distinct clusters via PACX library using additional demons

PACX uses on each machine the individual, "native" MPI library for communications within one machine. This makes sense since the MPI libraries are normally highly tuned for each hardware platform. Across machines, PACX handles MPI calls via TCP connections.

5 Future User Perspective

The main users' perception is the steady increase in complexity. Researchers are no experts in all fields contained in a coupled simulations, and often want to use modules as black box without looking into the details. Interdisciplinary work is becoming more and more important in daily work. It requires supporting collaborative environments where people located on different sites, having different knowledge of modules as well as coding, computer science and high performance computing. Tools include workflow tools, portals, middleware or collaboration tools such that researchers can concentrate on their own fields of interest. Coupling tools are being developed either on top of individual codes, i.e. able to couple codes without insight in the source code, or within a single, strongly coupled code, i.e. as a library.

6 Requirements, Architecture and Tools

From the users perspective on future applications and daily work context, a list of necessary requirements on future computing systems can be extracted. The main point is that various modules combined into one application impose different needs upon the architecture. Therefore the system should be hybrid to some extend to fulfill competing necessities. This is especially relevant when some modules are not available on the favored destination platform. For the fluid-structure interaction, for example, the costly fluid part should, for performance and efficiency reasons, run on a supercomputer whereas the commercial structure simulation code is only available on standard PC clusters. Hardware developments either aim at integrated systems or HPC systems tightly connected to standard platforms. Software requirements range from coupled MPI libraries for applications over global file systems to enable or ease pre- and post-processing of distributed simulations to co- or conditional scheduling in a heterogeneous environment. Furthermore, hybrid debuggers and hybrid performance analysis tools are needed, and workflow tools and/or meta-languages are necessary in order to ease the distributed, interdisciplinary work.

Acknowledgements The experiences described in this paper were obtained during the Teraflop Workbench project and many other projects carried out at HLRS in cooperation with users of our systems. I want to thank all contributors, cited and uncited, for the fruitful discussions and overviews over their every day simulations and requirements.

References

1. Auweter-Kurtz, M., Fertig, M., Petkow, D., Stindl, T., Quandt, M., Munz, C.D., Adamis, P., Resch, M., Roller, S., D'Andrea, D., Schneider, R.: Development of a hybrid PIC/DSMC code. IEPC-2005-71, Proceedings of the 29th International Electric Propulsion Conference, Princeton, USA (2005)
2. D'Andrea, D., Quandt, M., Petkow., D., Stindl, T., Fertig, M., Auweter-Kurtz, M., Munz, C.D., Roller, S., Adamidis, P., Schneider, R.: Numerical simulation of instationary magnetoplasmadynamic thrusters. ESAIM: proceedings of CEMRACS (2005)
3. Keller, R., Liebing, M.: Using PACX-MPI in metacomputing applications. In: 18th Symposium Simulationstechnique, Erlangen (2005)
4. Klimach, H., Roller, S., Munz, C.D.: Heterogeneous parallel aero-acoustics using PACX-MPI. In: High-Performance Computing on Vector Systems 2008. Springer (2008). To be published
5. Lippold, F., Ruprecht, A.: Fluid-structure interaction in turbine simulation. In: High-Performance Computing on Vector Systems 2008. Springer (2008). To be published
6. Petkow, D., Fertig, M., Stindl, T., Auweter-Kurtz, M., Quandt, M., Munz, C.D., Roller, S., D'Andrea, D., Schneider, R.: Development of a 3-dimensional, time accurate particle method for rarefied plasma flows. AIAA-2006-3601, Proceedings of the 9th AIAA/ASME Joint Thermophysics and Heat Transfer Conference, San Francisco, USA (2006)
7. Scheibe, A., Westkämper, E.: Complex numerical simulations in production techniques. In: High-Performance Computing on Vector Systems 2008. Springer (2008). To be published

III
Grid Computing & Data Analysis

An Application of the NAREGI Grid Middleware to a Nationwide Joint-Use Environment for Computing

Eisaku Sakane, Manabu Higashida, and Shinji Shimojo

Abstract In the Cybermedia Center, Osaka University (CMC), we have applied the NAREGI grid middleware to our latest large-scale computing system for nationwide joint-use, and have made grid computing technology available on them coexisting with the conventional computing style. In this paper, we mention the problems of resource provisioning, password/passphrase management, issuance of grid certificate and grid operations in user management, and a solution is discussed.

1 Introduction

There are many grid systems in e-Science research. For example, for scientists to access and analyze data produced from the LHC (Large Hadron Collider), CERN is building a distributed computing and data storage infrastructure, the LHC Computing Grid (LCG) [3]. The TeraGrid [8] provides a scientific discovery infrastructure in USA, so that researchers can access high-performance computers and data resources. The PRAGMA (Pacific Rim Application and Grid Middleware Assembly) [6] advances the use of grid technologies in applications among a community of investigators working with leading institutions around the Pacific Rim. For international collaborations, grid technology becomes essential.

In Japan, the NAREGI (National Research Grid Initiative) project [5] began in 2003. It makes fundamental building blocks in the Cyber Science Infrastructure (CSI) project promoted by the National Institute of Informatics (NII), and its goal is to provide an integrated grid computing environment for widely-distributed re-

Eisaku Sakane · Manabu Higashida
Cybermedia Center, Osaka University, 5-1 Mihogaoka, Ibaraki, Osaka 567-0047, Japan, e-mail: {sakane, manabu}@cmc.osaka-u.ac.jp

Shinji Shimojo
National Institute of Information and Communications Technology, KDDI Otemachi Build. 21F, 1-8-1 Otemachi, Chiyoda-ku, Tokyo 100-0004, Japan, e-mail: sshinji@nict.go.jp

search and education (the Science Grid). The NAREGI grid middleware has been developed in order to create such an environment.

As a nationwide joint-use facility, CMC provides academic researchers with a large-scale computing system consisting of 20 nodes of vector-type supercomputer NEC SX-8R (5.3 TFlop/s, 3 TB memory in total) and cluster systems composed of more than 600 PCs (24.4 TFlop/s, 3 TB memory in total). To promote not only the e-Science in Japan but also the international collaborations, we have to introduce grid technologies to our High Performance Computing (HPC) system, to provide the user with our computational resources as grid resources and advance the use of grid computing.

Taking advantage of our HPC system update of the beginning of 2007, we have planned the construction of a grid computing environment on our large-scale computing system and have decided to introduce the NAREGI grid middleware $\beta 2$ co-existing with the conventional style of computing [7]. However, there are several problems in the application of the NAREGI middleware to our system.

After giving a brief survey of the NAREGI middleware in Sect. 2, we will mention several issues in the application of the NAREGI middleware $\beta 2$ in Sect. 3. A method solving the issues will be proposed in Sect. 4. In Sect. 5 and Sect. 6 the proposed system will be evaluated. Finally we will summarize in Sect. 7.

2 NAREGI Grid Middleware

The NAREGI grid middleware is developed by the Center for Grid Research and Development at NII and provides an integrated grid environment. The platform of the NAREGI middleware after β release is the Globus Toolkit 4 [9]. The NAREGI middleware functions are as follows: resource management, data-grid, security, user-environment, programming environment and application support. For later convenience, we mention the resource management, security and user environment of the NAREGI middleware and omit the others. A standard constitution of NAREGI middleware is shown in Fig. 1.

The NAREGI resource management is taken by the Super Scheduler (SS), the Information Service (IS) and the GridVM (Virtual Machine). The SS searches an available computational resource, reserves it, and then submits a job. The IS manages detailed information about each resource and provides them to the SS. The GridVM virtualizes computational resources, replies a request from the SS, and delivers a job to a local scheduler.

Authentication and authorization functions are provided for the implementation of the Grid Security Infrastructure (GSI). The authentication in the NAREGI is based on Public Key Infrastructure (PKI) in conformance with the X.509 certificate format and on the delegation mechanism of proxy certificates. The NAREGI-CA (Certification Authority software) makes it possible to build a grid CA.

In the NAREGI middleware $\beta 2$, a user's key pair together with a user certificate are managed in the User Management Server (UMS) [4], and are not stored

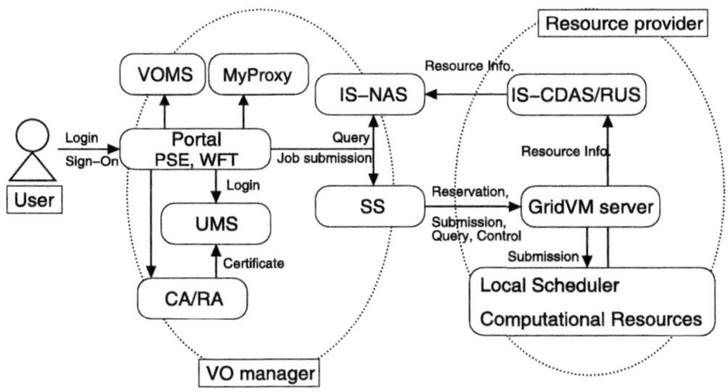

Fig. 1 NAREGI grid middleware

into the user terminal. In issuing a proxy certificate, it is signed by the certificate stored in the UMS through the NAREGI portal. The user must manage three passwords/passphrases: a password for the user ID on UMS, and passphrases for private keys of the grid user certificate as well as the proxy certificate.

In general, users and resource providers form a Virtual Organization (VO) on a grid network according to their access policies. The information about users and resources belonging into the VO can be used in an authorization process. The Virtual Organization Membership Service (VOMS) manages such VO information.

The NAREGI user environment is composed of three components, WFT (Work Flow Tool), GridPSE (Problem Solving Environment), and GVS (Grid Visualization System). Users can use these components as web services on the NAREGI portal.

3 Issues

In this section, we describe several issues in the application of the NAREGI middleware $\beta 2$ to our HPC system.

In our HPC system, before the introduction of the NAREGI middleware, a user usually logs in a front end server using SSH and submits a job to the batch queue system through the Command-line User Interface (CUI). We have to maintain the conventional style of computing because it is hard for all of users to shift quickly from the CUI to the NAREGI interface. Hereafter, we consider coexistence with the conventional style of computing to be a fundamental requirement.

3.1 Resource Provisioning

CMC has unified the scheduler of the vector-type supercomputer system (SUPER-UX) and the PC cluster system (Linux) by introducing the NEC NQS II batch processing system with JobManipulator which provides functions of a fair share scheduling and a backfill scheduling. This unification of schedulers has the following advantages: it provides a unified interface for both user and administrator regardless of resource architectures. Moreover, the user can perform easily a coupled simulation between different types of architectures.

Unfortunately, the NAREGI middleware $\beta 2$ does not support the NQS II for Linux. If we use a scheduler supported by the NAREGI middleware $\beta 2$, the benefits of unification mentioned above may be lost. Furthermore, new problems such as resources division, individual accounting system, etc., may be caused by the existence of plural schedulers.

3.2 Password / Passphrase Management

To integrate the authentication at each server into a strongly protected authentication server, CMC has introduced the Kerberos authentication system constructed by Microsoft Active Directory server and has realized Single Sign-On (SSO) access to each server. Because of GSSAPI (IETF/RFC4462), the user can log in a terminal server using SSH with the Kerberos authentication. By the SPNEGO mechanism negotiation (IETF/RFC4559), the combination of almost all client OSs and web browsers, which are in circulation at present, realizes SSO to each server in our HPC system.

The NAREGI middleware $\beta 2$ cannot use the Kerberos authentication system we have introduced. Therefore, two authentication systems will exist independently in our HPC system if we construct a portal of the grid computing by just using the NAREGI middleware $\beta 2$. This forces a user to manage multiple IDs and passwords for both authentication systems. Proper management of such password/passphrases is a burden to the user. Also, such a system is easy to attack with social engineering.

3.3 Issuance of Grid Certificate

For promoting grid technology, CMC gets infrastructure ready so that all users can use the grid environment. It is necessary for all users to have a grid certificate. A reliable grid certificate is necessary to cooperate with not only the domestic organizations but also the overseas organizations. The issuance of a grid certificate requires personal identification through interview regarding the PMA's (Policy Management Authority) classic profile.

So far, the application procedure for the use of computing resources has not required personal authentication. It is difficult to let all users apply for a certificate issuance at a CA[1] authorized by the PMA, separately from the ordinary application as user for our center. Even if the user obtained the grid certificate, a method to store away the private key to the UMS securely must be established.

3.4 Grid Operations

In order for a user to use the NAREGI system, the following administrator's operations are needed in addition to the conventional operations: certificate issuance preparations, VO information registration, and adding the user entry to the grid-mapfile on the computational resources (GridVM). These grid operations are performed on each node and will be hard if the number of grid users is increase. Unfortunately, the NAREGI middleware $\beta 2$ does not provide a component supporting such maintainer's operations.

4 Application Method

In this section, we propose an application method solving the issues mentioned in previous section.

Figure 2 shows the constitution of the proposed HPC system. For conventional style of computing, a user gets a Kerberos ticket form Kerberos KDC (Key Distribution Center) and logs in the Login server with the Kerberos credential. The front end server provides the user with a terminal working environment for job submission.

To solve the issue described in Sect. 3.1, it is natural to make the GridVM support the NQS II for Linux. Thus, CMC develops the GridVM supporting the NQS II for Linux. One of GridVM servers in Fig. 2 is for the vector-type supercomputer and the remainder is for PC cluster systems.

To simplify password/passphrase management, CMC improves the NAREGI portal to enable a user to log into the UMS with the Kerberos ticket. The NAREGI portal is constructed with Apache web server and Tomcat, which uses `mod_jk2` as Tomcat Connector. For Kerberos authentication on Apache, the Kerberos module for Apache (`mod_auth_kerb`) is installed into the Portal and RA servers respectively.

Furthermore we decided to automate certificate issuance procedure and remove passphrase inputs from the procedure. The hashed characters generated uniquely by combining a UserID and a certain salt are used to set as the passphrases of private

[1] In Japan, AIST GRID CA, KEK GRID CA, and NAREGI CA are authorized by the APGrid PMA [2] Production Level CA as of November, 2007. AIST GRID CA operated by the Grid Technology Research Center of National Institute of Advanced Industrial Science and Technology (AIST). KEK GRID CA operated by KEK (High Energy Accelerator Research Organization) Computing Research Center.

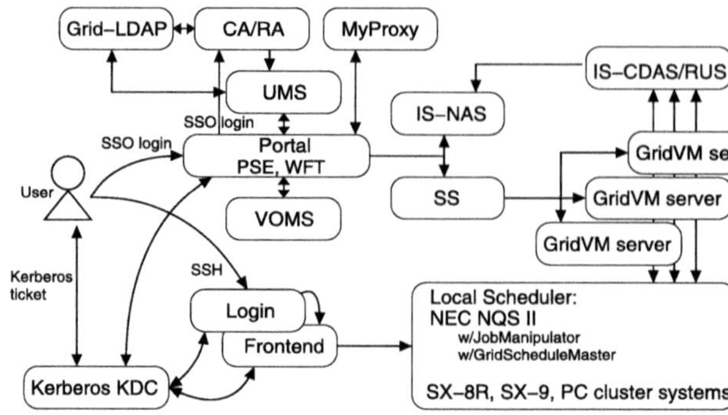

Fig. 2 CMC HPC system constitution

keys. To automate the issuance procedure, it is necessary to prepare beforehand a Subject DN of the certificate and a LicenseID that is used at the requesting RA to sign. Then, we build a LDAP server (Grid-LDAP in Fig. 2) to manage such information about users.

For the problem of issuance of grid certificates, we give first priority to the issuance of grid certificates to all nationwide users of our center and construct a grid CA at CMC. For the construction of CA, we follow the classic profile of APGrid PMA[2] as much as possible and stipulate a CP/CPS (Certificate Policy and Certification Practice Statements) in which person identification is not imposed. According to the minimum CA requirements by the APGrid PMA, the passphrase of the user's private key must be more than 12 characters. The strength of the passphrase equals to that of the Kerberos password because the proposed system sets the passphrase of the private key automatically. Thus, we impose that Kerberos password must be at least 12 characters long.

To promote efficiency of grid operations, especially user management operations, we develop a user management system which automates grid operations on each node and works together with the conventional user registration system. The core component chart of system is shown in Fig. 3.

The user management system, Navias, creates a user account on the UMS and the GridVMs, registers VO information, e.g., VO name and role, and adds a user entry to the grid-mapfile on the GridVMs. The system is composed of a controller, agents on each server, and the Grid-LDAP. The controller works together with the conventional user registration system, refers to the Grid-LDAP server for necessary user information, and send requests to the agents. We manage all necessary user information in the Grid-LDAP.

By the specification of the NAREGI, users and resources must belong to a certain VO. We have to decide about the constitution of VO in order for all users at our center to use the NAREGI system. We form the VO called CMC_Osaka as follows; VO users are all of users at our center. Resource provider and VOMS manager is CMC. Such a VO constitution is elementary for a nationwide joint-use facility.

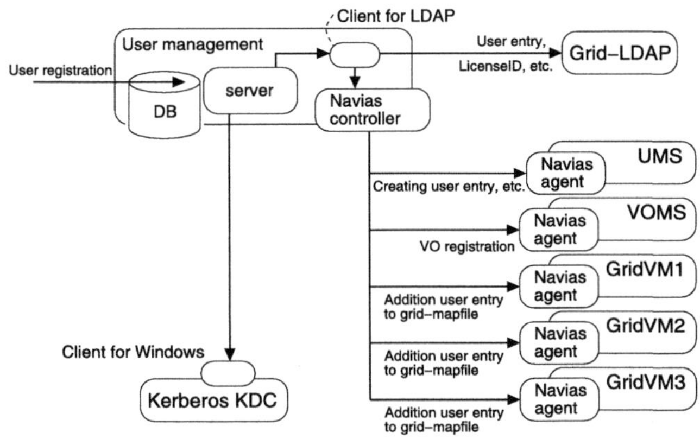

Fig. 3 Core components of Navias

5 Evaluation of Proposed Method

With the improvement of GridVM to support NQS II, all jobs can be handled by NQS II regardless of its nature (grid job or non-grid job). This means that CMC will be able to provide computing resources fairly for users according to a fixed amount accounting system based on the fair share scheduling using NQS II. For deployment of grid technology, the NAREGI middleware should support many schedulers. In this sense, our development contributes to the further propagation of grid technology.

On the authentication framework, a user can access both of the NAREGI portal and the conventional front end server by using only Kerberos ID and password. The automatically issuance function reduces the total number of password/passphrase managed by user from 4 to 1. As the user certificate is stored in the UMS, the security level of the system does not go down because our system reduces the cracking with social engineering[2]. Moreover, the proposed system simplifies the sign-on procedure to use the grid tool, thus usability in signing-on the NAREGI portal improved greatly.

Regarding the user certificate issuance, the NAREGI-CA software uses in general a License ID for user authentication through web enroll. However, such authentication by License ID is unnecessary in our proposal because the user is able to access the web enroll server with the Kerberos credential. License IDs are used to control the issuance number of user certificates that are issued according to those purposes.

Our CA satisfies the minimum CA requirements by the APGrid PMA except for the identification with an interview. The certificate issued by our CA cannot be commonly used in a PKI domain of the CA according to PMA's classic profile. However,

[2] In the case that a private key is stored in a device like an IC card, even if a passphrase is stolen, the account is not used unjustly unless the IC card is stolen.

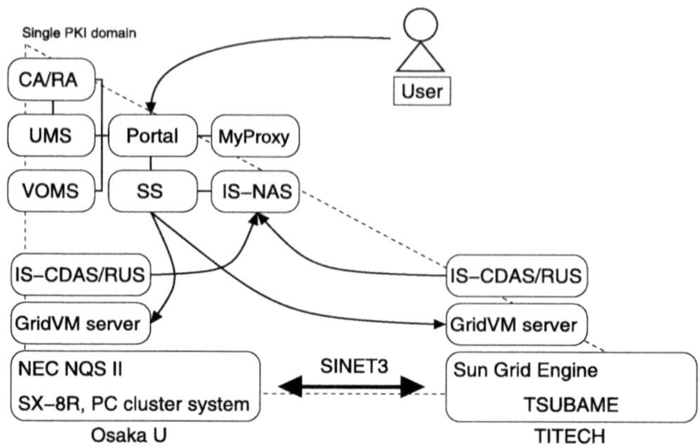

Fig. 4 Node composition chart at the first phase

our CP/CPS is still suitable for a nationwide joint-use environment for computing, because the supercomputer centers of the seven major national universities in Japan do not require personal identification with interview in the application procedure for the use of computing resources.

Currently, during the registration by the user management system, the VO name is fixed to CMC_Osaka to which the users of our center belong by default. Multiple VOs cannot be handled by the system. We must improve the system for multiple VO support.

6 Cooperative Evaluation

Toward realization of the Science Grid, CMC, Tokyo Institute of Technology (TITECH), Kyushu University and the NAREGI project started a cooperative evaluation experiment in April 2007. As first phase, we constructed the grid environment as shown in Fig. 4. Our HPC system and the TSUBAME system of TITECH was connected by SINET3 which is a dynamic L1/L2/L3 network provisioning a 40 Gbps backbone. The VO, called CMCGSIC_Osaka, was organized. In the CMCGSIC_Osaka, the VO manager was CMC, and the resource providers were CMC and TITECH. All of user and server certificates were issued by the grid CA at CMC. The grid system during the first phase was demonstrated at the SC07 Conference. We succeeded in executing jobs at all resources. Moreover, a large scale Molecular Dynamic simulation of the Institute of Laser Engineering at Osaka University was executed.

There were many problems in the evaluation during the first phase. Problems of the NAREGI middleware $\beta 2$ itself were reported to the NAREGI project. Here, we comment on problems with an operative aspect.

To use all computational resources in the grid environment, a user must perform the use application procedure for both CMC and TITECH because each center has an individual accounting system. Currently, a user must complete all user application procedures for each computing center (resource provider) participating in the VO unless the user uses a specific resource. This is a burden of the user. The user application procedures should be integrated. To do so, a cooperative accounting system will be needed between resource providers. A means of propagating securely the user information that each computing center needs should be established.

7 Summary and future works

In this paper, we have described the application of the NAREGI middleware $\beta 2$ to the nationwide joint-use environment for computing. In particular, the issues of resource provisioning, password/passphrase management, issuance of grid certificates and grid operation in user management have been discussed, and a method solving them has been proposed and evaluated.

On promoting cooperation with the other supercomputer centers and institutions, it is needed to form a VO flexibly. The supercomputer center should fill the role of hosting VOs in e-Science. To solve problems mentioned in Sect. 6 we aim at improvements of usability, grid operations and interoperability. We will report the results of the cooperative evaluation project in near future.

In Osaka University, a Campus-wide IT Authentication Platform System [1] has been developed. It is important to consider the trust federation between Campus and Grid PKI domains. The proposed system will able to support the trust federation with Campus PKI domain by using PKINIT (IETF/RFC4556 and 4557) that uses PKI for initial authentication in Kerberos.

Acknowledgements The authors would like to thank the members of the NAREGI project funded by MEXT (the Ministry of Education, Culture, Sports, Science and Technology) for their support. The authors would also like to thank the researchers of the Institute of Laser Engineering at Osaka University for their cooperation. This work is partly based on the Research and Development Program for Cyber Science Infrastructure (CSI) promoted by the National Institute of Informatics, Japan.

References

1. Akiyama, T., Teranishi, Y., Okamura, S., Sakane, E., Hasegawa, G., Baba, K., Nakano, H., Shimojo, S., Nagaoka, T.: Campus-wide IT Authentication Infrastructure Development in Osaka University. IPSJ Journal **49**(3), 1249–1264 (2008)

2. APGrid PMA. http://www.apgridpma.org/
3. LCG project. http://lcg.web.cern.ch/LCG/
4. Mineo, S., Yotsu, T., Saga, K., Saeki, Y., Tanaka, Y.: A Security Model in Grid Computing: An Implementation in the NAREGI project. In: IEICE Technical Report IA2006-21, pp. 7–12 (2006)
5. NAREGI project. http://www.naregi.org/
6. PRAGMA. http://www.pragma-grid.net/
7. Sakane, E., Higashida, M., Okamura, S., Teranishi, Y., Akiyama, T., Baba, K., Shimojo, S.: An application of grid middleware to a nationwide joint-use environment for computing. In: IPSJ SIG Technical Report 2007-DSM-45, pp. 25–30 (2007)
8. TeraGrid. http://www.teragrid.org/
9. The Globus Toolkit Homepage. http://www.globus.org/toolkit/

Interoperation between Atomic Energy Grid Infrastructure (AEGIS) and Other Grids

Yoshio Suzuki, Noriyuki Kushida, Naoya Teshima, Kohei Nakajima, Akemi Nishida, and Norihiro Nakajima

Abstract Coordination of global knowledge is needed to advance the computational and computer science needed for nuclear research. We have been conducting cooperative international research in various fields to construct a highly-secure worldwide network computing infrastructure, based on the Atomic Energy Grid Infrastructure (AEGIS). A promising way to achieve this is to establish interoperation using AEGIS with other grids. The operation of existing grid environments that allow the continuous development and execution of user applications is critical. To achieve the interoperability while maintaining operations, we have developed a system that converts messages among different grid middlewares without requiring modification of grid middlewares. To realize interoperability with two or more grid middlewares in the present study, we have defined the application programming interface (API) as a common interface to convert messages among the grid middlewares. We have applied our system to three interoperable environments. Through these experiments, we have confirmed that our system is applicable to the construction of interoperable environments among various grid middlewares.

1 Introduction

Coordination of global knowledge is needed to advance the computational and computer science needed for nuclear research. The construction of a worldwide computing infrastructure network that would share and integrate worldwide computer resources, tools and data with high security is an important step toward realizing this goal. For example, resources must be shared and integrated effectively to en-

Yoshio Suzuki · Noriyuki Kushida · Naoya Teshima · Kohei Nakajima · Akemi Nishida · Norihiro Nakajima
Center for computational science and e-systems, Japan Atomic Energy Agency, 6-9-3 Higashi-Ueno, Taito-ku, Tokyo 110-0015, Japan, e-mail: {suzuki.yoshio, kushida.noriyuki, teshima.naoya, nakajima.kohei, nishida.akemi, nakajima.norihiro}@jaea.go.jp

able the execution of large-scale simulations, coupling simulations or parametric simulations, which are impossible or difficult to perform using only one computer in each institute or university. Such simulations are indispensable to the nuclear research field. Promising technology is being developed that would allow interoperability using different grid middlewares to achieve a worldwide network computing infrastructure.

Research on the grid computing technology began approximately ten years ago and has resulted in the development of various grid middlewares throughout the world. To date, research concerning the interoperability between different grid middlewares has been promoted under the international programs of grid middleware standardization. One example is UniGrids [2], which was funded under the 6th framework Specific Targeted Research Project (STREP) and was carried out for two years beginning in July 2004. The goal of the project was to attain the interoperability of the Uniform Interface to Computing Resources (UNICORE) system [1] and Globus [4]. The project enabled UNICORE users to utilize computers managed by Globus.

In the current paper, we describe our research and development (R&D) of the interoperation technology using the Atomic Energy Grid Infrastructure (AEGIS) [8, 7] among other grids and three actual experiments executed under three international collaborations.

2 History of Grid Computing Technology at CCSE

The center for computational science and e-systems of the Japan Atomic Energy Agency (CCSE/JAEA) leads in computer science and computational science R&D, aiming to predict phenomena in nuclear science and engineering. As a cutting edge nuclear science and engineering organization, we focus on fusion, fission and proton accelerator research. In support of this research, CCSE carries out computational science and computer science R&D as well as operation and maintenance of the supercomputers and networks.

CCSE has been conducting grid computing technology R&D since the fiscal year (FY) 1995, working toward establishing an infrastructure for computational science research. The R&D effort was divided into four projects (see Fig. 1).

R&D of the Seamless Thinking Aid (STA) project [10] was undertaken first, from FY1995 to FY2000. As a result, an environment in which researchers and engineers can seamlessly use distributed heterogeneous computers has been established. Protocols have been developed to implement STA, including a communication library to execute remote procedure calls (STA Remote Procedure Call (STARPC)), an environment to develop programs such as editors, compilers, debuggers and workflow tools to control job submissions and data transfer. Using computers under the STA environment, improved the throughput of a coupling simulation from 30 to 70% compared to the use of a single computer. Using this technology, a simulation that utilized 510 CPUs distributed among five institutes in Japan, Europe and the US was executed at SC2000, the High Performance Networking and Computing Conference.

The second project was the R&D of the Information Technology Based Laboratory (ITBL) infrastructure [5, 9]. The ITBL project is a part of the national E-Japan National Priority Policy Program, developed to realize the e-Japan Strategy goals of making Japan the world's most advanced IT nation. ITBL seeks to support a virtual laboratory for large-scale computational science and engineering simulations by facilitating interactions among heterogeneous computing resources, information systems and instruments, all of which are geographically and organizationally dispersed. The ITBL project was launched in FY2001 by six institutes: the National Institute for Materials Science (NIMS), the National Research Institute for Earth Science and Disaster Prevention (NIED), the Japan Aerospace Exploration Agency (JAXA), the Institute of Physical and Chemical Research (known as RIKEN), the Japan Science and Technology Agency (JST) and JAEA, sponsored by the Ministry of Education, Culture, Sports, Science and Technology (MEXT) for five years. Under the ITBL project, CCSE developed the ITBL infrastructure by expanding STA technology.

Figure 2 shows the principal functions of ITBL infrastructure software. The functions include authentication by X.509, workflow, communication and large-scale visualization. These functions can be used seamlessly by institutes and universities connected through the Internet. Encrypted communication using HTTPS protocols maintains high security on the Internet and enables communications to coexist with firewalls. We conducted R&D of the infrastructure software by considering the security policies of the computer and network managements at Japanese institutes and universities. As a result, the ITBL infrastructure software was able to connect 23 computers (including 21 different models) at 12 institutes and universities by the end of FY2007. The total peak performance is up to 73 TFlop/s. Figure 3 shows the established ITBL environment. Since the conclusion of the project, the ITBL environment has been continuously managed through an extended arrangement. By the

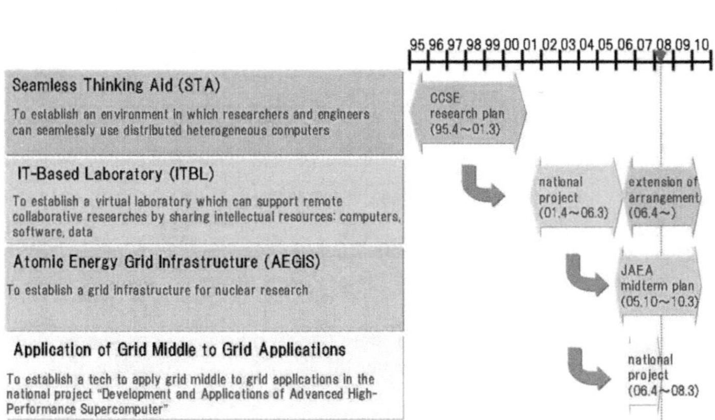

Fig. 1 Grid Computing Technology R&D at CCSE

end of FY2007, approximately 800 users from 128 organizations were participating in the ITBL. Users have generated 13 user communities. We have developed the ITBL client application programming interface (API) to satisfy user demand to develop a client system that enables their control of the grid-enabled application from a client terminal.

The third project was the R&D of AEGIS, which was included in the JAEA midterm plan. The objective was to establish a grid infrastructure for nuclear research. Details of the AEGIS project are described in Sect. 3.

The last R&D project involved the utilization of grid middlewares for applications [7], which was included as part of the national Development and Applications of Advanced High-Performance Supercomputers project and is described in Sect. 4. Under this phase, we established APIs to utilize grid middlewares for applications.

3 Atomic Energy Grid Infrastructure

The objective of AEGIS was to develop computational infrastructure for atomic energy research. The R&D aimed to establish predictable science and engineering to guarantee security and safety of atomic energy and to preserve the environment by constructing a real-scale nuclear facility using computer-aided R&D, engineering and science as well as a network computing system that enables their synchronicity. To develop the base of AEGIS, a network computing systems was constructed by expanding on the knowledge and skills of ITBL infrastructure software.

The safety, certainty and reliability of large-scale nuclear facilities and supercomputers for atomic energy research needed to be maintained for the Internet and the communication infrastructure developed in ITBL was improved for increased

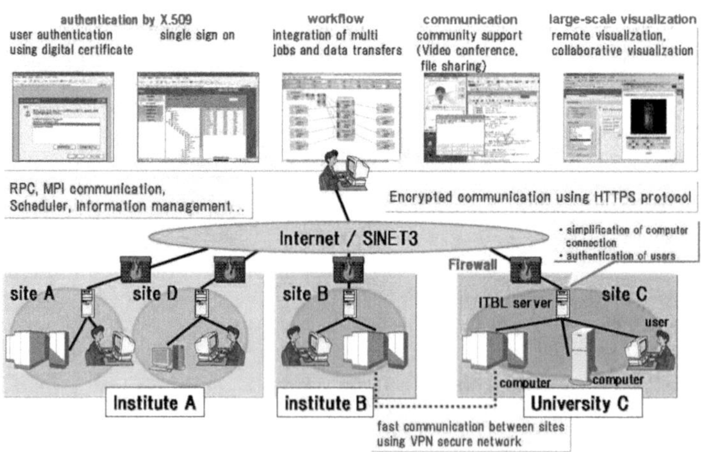

Fig. 2 Principal functions of ITBL infrastructure

Internet communication security using more reliable communication middlewares. Currently, AEGIS is being used as the communication infrastructure for remote experiments in nuclear fusion research. The communication infrastructure has also been used for a three-dimensional virtual plant vibration simulator. To improve the security of accessing JAEA's Local Area Network (LAN) from the Internet, we implemented the use of a digital certificate and have installed file servers and video conferencing systems. The system allows certified users to share information.

4 Interoperability Technology based on Atomic Energy Grid Infrastructure

Collaborative R&D among Japanese institutes and universities and German, French and U.S. nuclear and computational agencies has furthered the development of computational science and information technology in the atomic energy research field. The German, French and U.S. agencies utilize unique grid middlewares and our R&D for the interoperation of these grids using AEGIS has allowed successful interoperability trials with institutes and universities in participating countries.

As described in the introduction, the UniGrids project enabled UNICORE users to utilize computers managed by Globus. However, the standardization of these systems demands their modification. The establishment of standardized systems enables easier grid middleware interoperation. Under standardization, each component of a different grid middleware system can cooperate more flexibly with another, enhancing the grid middleware. Consequently, the grid middleware systems that have already been constructed by adopting various architectures worldwide require modification. Although the operation of grid environments that allow the continuous development and execution of their applications is critical, the alterations under

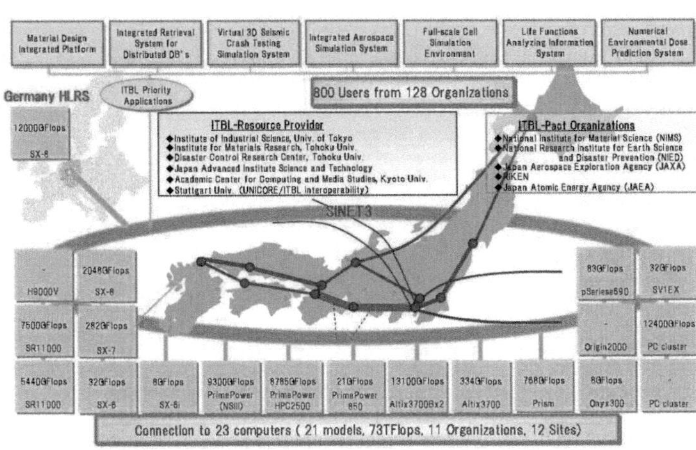

Fig. 3 Established ITBL environment

standardization prevent users from continuing the development of their applications on grid environments. Interoperability would facilitate the concurrent development of new applications by different users. One of the simplest ways to achieve interoperability while maintaining operations is to convert messages among different grid middlewares without modifying their content [9]. Furthermore, in order to realize interoperability with two or more grid middlewares, a common interface that converts messages for those grid middlewares must be defined [7].

In the present study, we focused on the ITBL client API. The client API is used to call grid middleware functions (ITBL infrastructure software) from their client systems. The API should also be enabled to call these functions from other grids. Therefore, we surveyed the use of the grid environment by the grid-enabled ITBL project applications, defined client API for AEGIS by expanding the ITBL client API and implemented the API modifications needed for grid interoperability. The classifications of grid-enabled applications for the ITBL project are shown on the left side of Fig. 4. The ITBL project applications are classified into four categories: parameter survey, computer substitution, combination of different architectures and virtual scalable parallel computing. Based on the classifications, we have defined the client API hierarchy under AEGIS, as shown on the right side of Fig. 4. The API is stratified into three levels: low, middle and high levels. The high level API includes functions corresponding to ITBL client systems. The high level API also includes functions from the perspective of application fields. The middle level API includes functions to construct the high level API and is further divided into three levels: the method level, integrated level and independent level. The low level API includes basic functions: authentication and communication. Using this AEGIS client API, we have constructed interoperable environments between AEGIS and other grid sites.

Figure 5 shows a simplified configuration of an interoperable system between AEGIS and other grid middlewares. We have implemented the AEGIS client API with other grids, including UNICORE, DIET [3] and Globus. Thus, the AEGIS

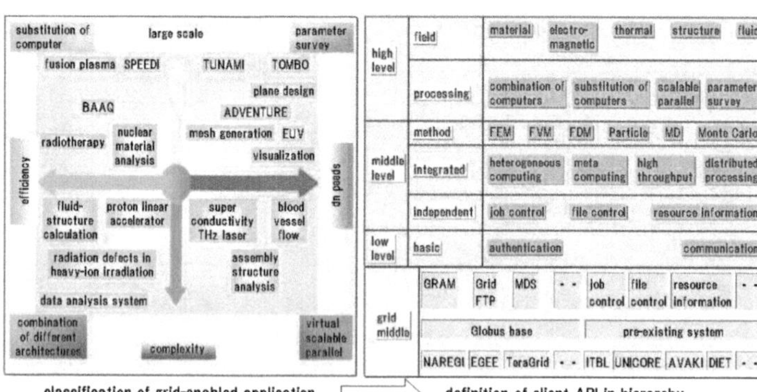

Fig. 4 Classification of grid-enabled ITBL project applications and the definition of client API hierarchy under AEGIS

client API generates common interfaces to call other grid middlewares. Here, authentication is the critical issue. When operating other girds from AEGIS, security is maintained by keeping a key for other grids on the AEGIS server with high security and authenticating a user using the authentication function of the low level API. A client system on a user PC calls the AEGIS middleware function using the client API. The message of the function is sent to the session manager and converted through the translator. The translator calls the corresponding function for each grid middleware through the client API which is implemented on each grid API. When operating AEGIS from other grids, the client API implemented in each grid middleware calls the AEGIS functions. By using this configuration, we have implemented interoperable systems between AEGIS and systems such as UNICORE, DIET and Globus. We have conducted cooperative international research in various fields to advance computer science and to extend the AEGIS grid computing environment as shown in Fig. 6. Using the preceding method, we have established interoperable environments between CCSE and each institute and university.

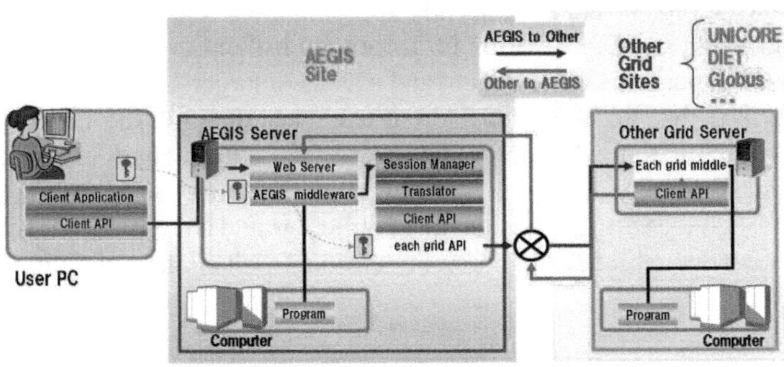

Fig. 5 Simplified configuration of an interoperable system between AEGIS and other grid middlewares

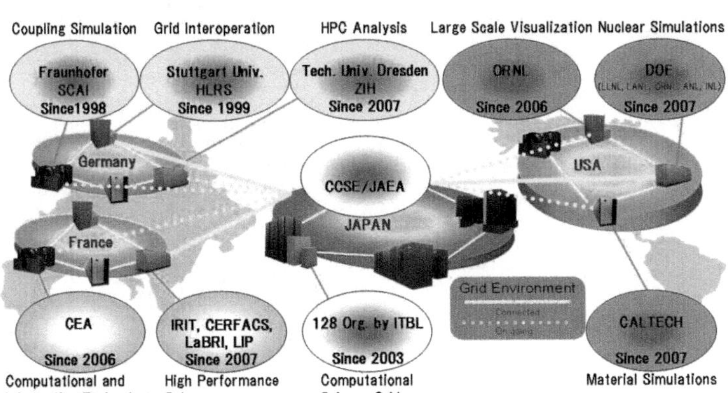

Fig. 6 Cooperative international research with CCSE/JAEA

5 Collaboration with the High Performance Computing Center Stuttgart (HLRS)

The objective of AEGIS is international cooperation in computational science and engineering R&D. Actions completed included: 1) the enhancement of tools for metacomputing, particularly in the area of performance analysis; 2) the enhancement of metacomputer applicable applications, including nuclear engineering simulation; and, 3) analysis of low-level network protocols and evaluation of quality of service (QoS).

The interoperable environment between AEGIS and UNICORE was implemented at HLRS to establish an environment that can share and integrate computer resources and execute a large-scale simulation between CCSE and HLRS. As a result, the NEC SX-8 installation at HLRS can be used from CCSE. Namely, it is possible to control jobs and files from computers in AEGIS and UNICORE and acquire the resource information stored on computers in AEGIS and UNICORE. Using the environment, we successfully executed the three-dimensional virtual plant vibration simulation (see Fig. 7) and are currently applying the interoperability technology at Fraunhofer SCAI and Das Zentrum für Informationsdienste und Hochleistungsrechnen (ZIH) to expand the environment and execute a larger-scale simulation.

To simulate the vibration behavior of a nuclear facility, the input — seismic waves propagating to the buildings and equipment — must be accurately evaluated. Such assessments require an understanding of the dynamics of seismic waves propagating through the earth's crust, the plant buildings and equipment. To achieve this, CCSE has conducted R&D of a three-dimensional virtual plant vibration simulator.

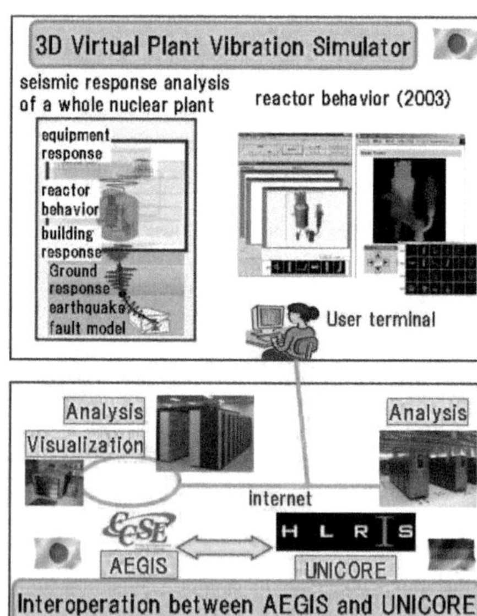

Fig. 7 System configuration of previous HLRS collaboration results

The grid-enabled application for elastostatic analysis was developed using the AEGIS client API. Figure 8 shows the established application and its workflow. The elastostatic analysis includes three solvers: CONNECT, the FInite Element STructural analysis for Assembly (FIESTA) and PartFIESTA. FIESTA, which was developed by CCSE for structural analysis of nuclear power plants as assemblages of parts, computes parts and components in parallel and individually on distributed computers, considering the functionality of each part. Each part solved in distributed computers is gathered to form an entire structure. The CONNECT solver calculates joint elements of parts from input files (parts mesh, parts interface and adjacent parts). PartFIESTA generates a domain-decomposed mesh. Using these data, the FIESTA solver executes structural analysis. Finally, output files are generated as visualization images. Users can control these files and executions using the grid-enabled application on the user PC.

In the current analysis, the High Temperature engineering Test Reactor (HTTR) was digitalized by parts and forms assembled an entire nuclear plant. The HTTR plant is composed of a reactor pressure vessel, cooling system components and pipes interconnecting these major components. We have successfully simulated the time evolution of an entire nuclear plant using the interoperable environment. However, simulated parts using the NEC SX-8 are small enough to avoid a communication bottleneck. A faster communication mechanism is critical for the execution of a large-scale simulation that uses all computer resources efficiently. To overcome this, we plan to research a new communication mechanism as part of the international collaboration.

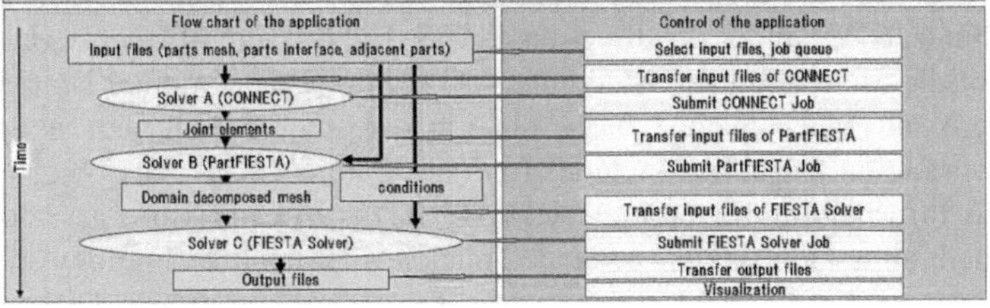

Fig. 8 Established application for elastostatic analysis and its workflow

6 Collaboration with Grid-TLSE Project Partners

Grid-TLSE project partners include the Institut de Recherche en Informatique de Toulouse (IRIT), the Centre Européen de Recherche et de Formation Avancée en Calcul Scientifique (CERFACS), the Laboratoire d'Informatique du Parallélisme (LIP) and the Laboratoire Bordelais de Recherche en Informatique (LaBRI). Grid-TLSE allows the comparative analysis of a number of direct solvers (free or commercially distributed) on user-submitted problems as well as on matrices from collections available on the TLSE website.

The objective of the collaboration with the Grid-TLSE project partners is cooperation concerning R&D of tools for high performance sparse linear algebra. Actions under the collaboration include: 1) cooperation on sparse linear solvers; and, 2) cooperation concerning AEGIS and GridRPC middleware DIET.

We aim to establish a sparse solver prediction system using computer resources at JAEA and Grid-TLSE project partners. Various sparse solvers have been developed in each institute worldwide and suitable pairings between sparse solvers and computers used to solve sparse matrices can be identified. To solve sparse matrices and use computer resources efficiently, it is important to find a suitable pairing. Currently, we have established a prototype of interoperable environments between AEGIS and DIET to execute the prediction system for the sparse solver Grid-TLSE [3]. In the prototype, the job execution from DIET to AEGIS is confirmed. Figure 9 shows the target system configuration. We plan to confirm the utility of the system using the FIESTA matrix.

7 Collaboration with the U.S. Department of Energy Under the Global Nuclear Energy Partnership

Six Working Groups (WGs) were organized as part of the Global Nuclear Energy Partnership (GNEP) under the direction of U.S. Department of Energy (DOE). The mission of the Simulation and Modeling WG (SMWG) is computer simulation and modeling of advanced nuclear fuels, materials, reactor design and validation using appropriate data. SMWG includes four sub-groups (SGs). The objective of the Advanced Information Technology SG (AITSG) is to establish a grid computing environment aimed to establish advanced nuclear fuel-cycle technology and to construct a fast reactor. Planned actions under the collaboration are: 1) the development of a network computing infrastructure to share tools and data between the U.S. and Japan aimed at verifying simulations and models; 2) to secure access to computer resources and to execute tools developed in the U.S. and Japan; and, 3) R&D of advanced technology in computer and information science needed to establish advanced fuel cycle technology and to construct a fast reactor. Figure 10 shows the schematic diagram of the themes espoused by AITSG. To establish the network computing infrastructure, the interoperation between AEGIS and Globus on the U.S. side needs to be effective. The computational science-aided laboratory on the network computing infrastructure plays an important role in accomplishing the themes established by the SMWG as well as the other working groups.

The sensitive nature of the tools and data utilized by GNEP require the highest security levels. Currently, we have implemented a prototype system for the network computing infrastructure. Under this prototype, it is possible to share tools and data on the interoperable environment between AEGIS and Globus while maintaining a high security level for each grid middleware.

8 Summary

R&D under AEGIS was conducted to provide a computational infrastructure for atomic energy research. International cooperation research in various fields has also been completed to advance computer science and to expand AEGIS.

In cooperation with HLRS, we successfully executed a three-dimensional virtual plant vibration simulation using the interoperable environment between AEGIS and UNICORE. The simulation confirms that job controls, file controls and acquisition of resource information are possible between computer resources of JAEA and HLRS using our interoperable system. As part of the cooperation with Grid-TLSE

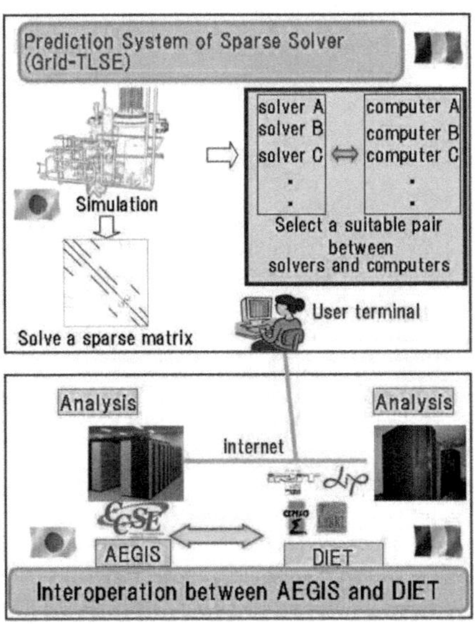

Fig. 9 Target system configuration used to execute the sparse solver prediction system under the interoperable environment between AEGIS and DIET

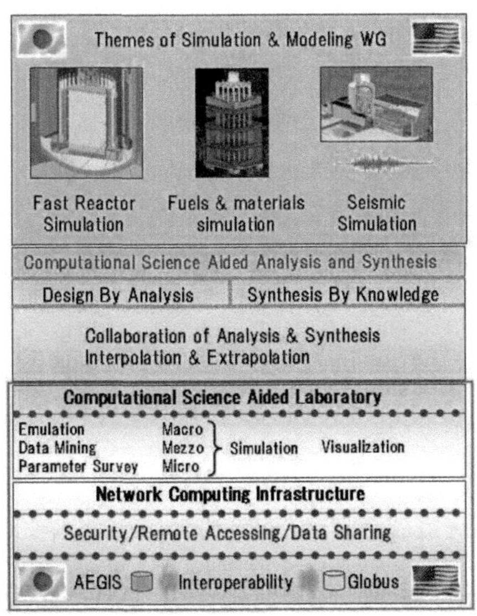

Fig. 10 Schematic diagram of the Simulation and Modeling Working Group themes and the contribution of the Advanced Information Technology Sub-Group

project partners, we established a prototype system for the interoperable environment between AEGIS and DIET to execute the sparse solver prediction system, confirming that job execution from AEGIS to DIET is possible using the prototype system.

Under GNEP, we collaborated with DOE to implement a prototype system for network computing infrastructure. The prototype maintains high security between computer resources and data files in AEGIS and Globus.

Based on these results, we confirmed that our concept of the interoperable system can be applied to construct an interoperable environment between AEGIS and other grids without modifying each grid middleware. Improvements to each prototype system will be needed to achieve research objectives as well as to advance the computational and computer science needed for nuclear research.

Acknowledgements The present study includes the results of two previous studies, the "Development and Applications of Advanced High-Performance Supercomputers" and the "Development of a Physical Model Describing the Dynamic Interaction Characteristics of Component Connections for the Analysis of an Entire Nuclear Facility" performed by the Japan Atomic Energy Agency and initiated by the Ministry of Education, Culture, Sports, Science and Technology of Japan as a result of the Strategic Japanese-French Cooperative Program on Information and Communications Technology, including Computer Science by the Japan Science and Technology Agency and the Centre National de la Recherche Scientifique. This research is partially carried out under international collaboration between the Center for Computational Science and Engineering/Japan Atomic Energy Agency (CCSE/JAEA) and the High Performance Computing Center Stuttgart, between CCSE/JAEA and the Institut de Recherche en Informatique de Toulouse, the Centre Européen de Recherche et de Formation Avancée en Calcul Scientifique, the Laboratoire d'Informatique du Parallélisme and the Laboratoire Bordelais de Recherche en Informatique and between CCSE/JAEA as well as the U.S. Department of Energy.

References

1. Almond, J., Snelling, D.: Unicore: Uniform access to supercomputing as an element of electronic commerce. Future Generation Computer Systems **613**, 1–10 (1999)
2. Brooke, J., Fellows, D., Maclaren, J.: Interoperability of Resource Description Across Grid Domain Boundaries. In: Proceedings of the European Congress on Computational Methods in Applied Sciences and Engineering (ECCOMAS 2004). Jyväskylä, Finland (2004)
3. Caron, E., Desprez, F.: DIET: A Scalable Toolbox to Build Network Enabled Servers on the Grid. International Journal of High Performance Computing Applications **20**(11), 335–352 (2006)
4. Foster, I., Kessseleman, C.: Globus: A Metacomputing Infrastructure Toolkit, International Journal of High Performance Computing Applications **11**(2), 115–128 (1997)
5. Higuchi, K., Imamura, T., Suzuki, Y., Shimizu, F., Machida, M., Otani, T., Hasegawa, Y., Yamagishi, N., Kimura, K., Aoyagi, T., Nakajima, N., Fukuda M., Yagawa, G.: Grid Computing Supporting System on ITBL Project. In: Veidenbaum et al. (eds.) 5th International Symposium on High Performance Computing (ISHPC2003), *LNCS*, vol. 2858, pp. 245–257. Tokyo-Odaiba, Japan (2003)
6. Kushida, N., Suzuki, Y., Teshima, N., Nakajima, N., Caniou, Y., Dayde, M.: Seamless connection of ITBL computers from Grid-TLSE toward international matrix predication system. In: Proceedings of VECPAR08. Toulouse, France (2008) (to be published)

7. Suzuki, Y., Nakajima, N., Araya, F., Hazama, O., Nishida, A., Kushida, N., Akutsu, T., Teshima, N., Nakajima, K., Kondo, M., Hayashi, S., Aoyagi, T.: Development of three-dimensional virtual plant vibration simulator on grid computing environment ITBL-IS/AEGIS, In: Proceedings of the 16th International Conference on Nuclear Engineering, 16th International Conference on Nuclear Engineering (ICONE16). Orlando, Florida. CD-ROM (2008)
8. Suzuki, Y., Nakajima, K., Kushida, N., Kino, C., Minami, T., Matsumoto, N., Aoyagi, T., Nakajima, N., Iba, K., Hayashi, N., Ozeki, T., Totsuka, T., Nakanishi H., Nagayama, Y.: Research and Development of Fusion Grid Infrastructure Based on Atomic Energy Grid InfraStructure (AEGIS). Sixth IAEA Technical Meeting on Control, Data Acquisition, and Remote Participation for Fusion Research, Fusion Engineering and Design **83**, 511–515 (2008)
9. Suzuki, Y., Minami, T., Tani, M., Nakajima, N., Keller R., Beisel, T.: Interoperability between UNICORE and ITBL. In: Proceedings of the Workshop on Computational Grids and Clusters (WCGC'2006), *LNCS*, vol. 4395, pp. 601–609. Rio de Janeiro, Brazil (2006)
10. Takemiya, H., Imamura, T., Koide, H.: Development of a software system (STA: Seamless Thinking Aid) for Distributed Parallel Scientific Computing. Joho Shori **40**(11), 1104–1109 (1999) (in Japanese)
11. Tani, M., Nakajima, N., Nishida, A., Suzuki, Y., Matsubara, H., Hazama, O., Kushida, N., Kawasaki, K.: A Methodology of Structural Analysis for Nuclear Power Plant Size of Assembly. In: Proceedings of the Joint International Topical Meeting on Mathematics & Computations and Supercomputing in Nuclear Applications (M & C+SNA 2007) (15–19 April 2007), pp.15–19. Monterey, CA (2007)

Parallel File Systems in European Grid Projects

Peter W. Haas and Michael M. Resch

Abstract High performance computing has gradually shifted from the realm of research into development and partially even into the production cycles of industry. High performance computers therefore have to be integrated into production environments that demand the simultaneous solution of multidisciplinary physics problems. Supercomputer centers can learn from these new challenges imposed by industry. The concepts of work flow and production cycle open up a new horizon for integrating systems and software into what is called a distributed "Teraflop-Workbench" approach. Tera- or rather exascale storage and communication infrastructures will be needed to support such an environment.

1 Introduction

Based on a long tradition in supercomputing at the University of Stuttgart, HLRS was founded in 1995 as a federal center for High-Performance Computing. HLRS serves researchers at universities and research laboratories in Germany and their external and industrial partners with high-end compute power for engineering and scientific applications.

Peter W. Haas · Michael M. Resch
HLRS, Nobelstr. 19, 70569 Stuttgart, Germany, e-mail: {haas, resch}@hlrs.de

2 The HLRS Framework

2.1 Target

Due to its embedding in the industrial environment of southwest Germany's high-technology region around Stuttgart, HLRS traditionally focuses on applications from the engineering sciences like computational fluid dynamics, combustion, structural mechanics, electromagnetics and process engineering. However, recently HLRS has extended its portfolio to medical applications, environmental applications and has started an initiative to bring new fields of application to the supercomputer.

The services of HLRS offer complete support to user groups such as physical modeling of parallel numerics and algorithms, the embedding of commercial packages and the visualization of results either remotely or in HLRS' virtual reality laboratory at Stuttgart.

Research groups at HLRS are involved in numerous projects targeting at the industrial use of HPC technology and at the further improvement of such technologies based on user feedback. This includes the participation in the standardization of base technologies.

2.2 Organization

HLRS is a central facility of the University of Stuttgart. Its responsibility is to support supercomputing at the national level and at the university. HLRS operates computing platforms together with T-Systems, T-Systems sfr and Porsche in a joint company named hww (Höchstleistungsrechner für Wissenschaft und Wirtschaft GmbH). The universities at Heidelberg and at Karlsruhe are also shareholders of this company. The purpose of this public-private partnership is the sharing of resources in order to benefit from synergies. This allows for a broader diversification of the available architectures. At the same time common funding allows for higher budgets which in turn allow the installation of larger systems. Both research and industry benefit from this. While it is the role of HLRS to provide access to all platforms for researchers, T-Systems does marketing for industry.

2.3 HLRS Computer Configuration

The HLRS/hww computer configuration is distributed over multiple campuses. It uses state of the art equipment in a highly secure environment. The major campuses are located at the universities of Stuttgart, Heidelberg and Karlsruhe as well as within the premises of Daimler AG, Deutsches Zentrum für Luft- und Raumfahrt (DLR), Porsche AG, and T-Systems GmbH. Most of the communication lines use dedicated fiber links with optical multiplexors that allow for an appropriate collection of link signals, e.g. Ethernet, Fibre Channel or InfiniBand. Transmission speeds typically range between one and forty Gbit/s.

The HLRS/hww compute systems provide a fair coverage of today's high performance computing architectures and implementations. The spectrum extends from Shared Memory Parallel systems (SMPs), like NEC Asama and Bull Novascale, via Parallel Clusters, like Cray Opteron and NEC Nocona, to Massively Parallel Processors (MPPs), like HP XC6000, and finally to Parallel Vector Processors (PVPs), such as NEC SX-6 and NEC SX-8. Individual system performance may range from 200 Gflop/s up to 12 Tflop/s.

The HLRS computer configuration is one of the three corner-pillars supporting the Ex@Grid framework.

3 Ex@Grid Framework

3.1 Gauss Center for Supercomputing

The Gauss Centre for Supercomputing, www.gcfs.eu, the alliance of the John von Neumann Institute of Computing (NIC), the Leibniz-Rechenzentrum (LRZ), and the Höchstleistungsrechenzentrum Stuttgart (HLRS), provides one of the largest and most powerful supercomputer infrastructures in Europe. The German Ministry of Science and Education (BMBF) has announced to support the development of Ex@Grid, a very high-speed data communication backbone between the three centers, by 30 million Euros. This is to promote the scientific co-operation between all major HPC sites in Germany and in particular between their user communities in the area of distributed HPC applications, virtual and augmented reality, storage systems and networking.

Ex@Grid wants to provide a complete all-optical, forward error corrected network for the high performance computing centers in Germany. This will enable new services, like distributed HPC workflows based on a common network-centric data management infrastructure, for all scientific users since the network services can be very low latency due to lack of contention as well as datagram loss.

3.2 Ex@Grid Design

In conventional packet communication networks we have learned to take the lowest cost switching option for express traffic depending on bandwidth and latency. Express traffic has been moved from "fat IP routers" to layer-2 switches over time. The next significant cost reduction will be achieved by optically bypassing very high rate express traffics on demand. Switching times for transparent optical paths are ≈ 1 ms.

Therefore, the design of Ex@Grid tries to follow the construction principles of intersection-free highways where the highways and feeders together form 3-dimensional add/drop systems. This can be perceived as a direct equivalent to a degree-3 ROADM system. In comparison however, optical transmission systems are not restricted with respect to the number of lanes. Up to ≈ 40 different wavelengths (colors) are feasible within the ITU-100 Grid, and each of those wavelengths may be used to carry a payload signal of up to 100 Gbit/s.

4 Consolidation in the Datacenter

4.1 Datacenter Ethernet

In future, it will be very important to restrict the number of link signals and communication protocols because scalable gateways between dissimilar networks are either very expensive or prone to loss of datagrams or other forms of information. The IEEE 802.3 standardization efforts toward the Datacenter Ethernet [1] all aim at the integration of LAN, SAN and IPC into a common Ethernet-based switching technology which will enable Quality of Service (QoS) via service-specific configuration parameters. The direct provisioning of transparent optical channels between HPC applications may turn out to be a very elegant solution from the users' perspective. Here, users may select the most appropriate communication protocols for the duration of a link provisioning in a way that is best suited to their application. Thus information loss due to transmission errors or blocking inside multi-stage switching networks will be avoided at large.

4.2 IEEE 802.3ar: Congestion Management

We have tried to compose a top-level view of the most important definitions of the IEEE P802.3ar Congestion Management Task Force [2] which have been collected from information available under http://www.ieee802.org/3/ar/public/index.html. First, the overview tries to address the essential differences amongst the established network cultures with respect to traffic type, preferred characteristics as well as the associated transmission latencies. Below that there are three categories, namely Parallel Links, Virtually partitioned Links and Flow control, which are supposed to enable the coexistence of today's three most prevalent traffic types — within a single Ethernet, finally. We aim at making Datacenter Ethernet technology available within Ex@Grid at a very early point in time (both within switching systems and at the local end system's interface). The final version of the IEEE 802.3ar standard is supposed to be available in September 2007. This should enable first industry products at the beginning of 2008.

5 Multicluster Parallel File Systems

5.1 Multicluster GPFS in DEISA

The Distributed European Infrastructure for Supercomputing Applications, DEISA, www.deisa.org, is a consortium of leading national supercomputing centers that currently deploys and operates a persistent, production quality, distributed supercomputing environment with continental scope. The purpose of this FP6 funded research infrastructure is to enable scientific discovery across a broad spectrum of science and technology, by enhancing and reinforcing European capabilities in the area of high performance computing. This becomes possible through a deep integration of existing national high-end platforms, tightly coupled by a dedicated network and supported by innovative system and grid software.

All high performance computing systems share data among the computing nodes with a Cluster File System, which offers users a single system data view and transparent data access. The extension of this data sharing model to a grid of geographically distributed HPC systems over a wide area network leads to the concept of a Global File System or Grid File System, which avoids data replication.

In a first step four DEISA sites, all running AIX systems, integrated their local IBM GPFS file systems via the 10 Gbit/s dedicated DEISA wide area network into a joint Grid File System, using newly added features in the GPFS software. The step from a homogeneous to a heterogeneous environment was performed by the integration of the PowerPC-Linux system at BSC as well as the SGI Altix systems at SARA and LRZ. Additional software will enable hierarchical storage management functionality on top of GPFS via TSM, thus conveying the perception of an infinite space distributed storage system.

5.2 Network-centered parallel HSM systems

The High Performance Storage System (HPSS) is a parallel data management software that provides both fast access to data as well as services for very large storage environments. It has been designed with high performance computers and data handling systems in mind and can easily adapt to the throughput and capacity increases required in this field [3, 7].

HPSS may be of interest in situations having present and future scalability requirements that are very demanding in terms of total storage capacity, file sizes, data rates, number of objects stored and number of users. The focus of HPSS is the network. Data storage and archiving is distributed across a high performance network with user selectable quality of service. Hence users may access their data randomly and directly irrespective of their hierarchical grouping. HPSS is one of the very few storage systems that allow control of the communication parameters as part of the service class concept.

HLRS has been actively working in the area of layering parallel file systems via extensions to Data Management APIs, e.g. the parallel version of XDSM DMAPI. This has helped to raise the acceptance level for interworking scalable parallel file systems at HPC sites. IBM has announced extensions to its GFPS and HPSS file systems that would allow for a layering of GPFS on HPSS under a common name space. Basically, user level agents will be placed on the front end file system (i.e. GPFS) that exploit HSM control information in order to automatically migrate or stage user data files or entire file objects, respectively.

At leading HPC sites, there has been a holistic view of the global file system idea for a next generation supercomputer center for some time. This idea was (and still is) that all storage requests should be directed towards a single, site-wide storage repository. However, a closer look to the technological implications of the computer memory hierarchy will reveal, that there is still sufficient need and justification for more than just a single parallel file system.

5.3 Projected mass storage system at HLRS

The projected mass storage system at HLRS [4], e.g., is going to implement a layering of GPFS on HPSS which will definitely suffice for the pre- and post-processing stages in our HPC workflow as well as for most of our scientific users. This is either by PNFS exports or GPFS mounts on the client systems. However, the NEC SX-8 parallel vector system will maintain its local NEC GFS production file system, and rather employ local HPSS data movers in user space in order to transfer HPSS files into the parallel vector's main memory and vice versa. Of course, this procedure greatly reduces interdependencies between compute and storage subsystems. Even more importantly, it helps to mitigate the huge discrepancies in performance that tend to develop over time between successive generations of HPC systems.

6 The Teraflop Workbench Concept

When HLRS started its latest request for proposals for an HPC platform it was clear from the beginning that the system offered would have to be part of a larger concept of supercomputing called the Stuttgart Teraflop Workbench [6, 5]. The basic concept foresees a central file system where all the data reside during the scientific or industrial workflow. A variety of systems of different performance and architecture are directly connected to the file system. Each of them can be used for special purpose activities. In general, one can distinguish between systems for pre-processing, high-performance computing and post-processing.

The concept is centered around NEC's Global File System (GFS). NEC SX-8, Asama (IA-64 shared memory system) and a cluster of Intel Nocona processors all have direct access to the same data via a Fibre Channel storage area network.

6.1 Workflow Example

Most users tend to prepare their mesh on one processor before they decompose it and transfer it to the parallel system. Given that the main memory of the core system is 9 TB, we decided for one node with large memory to be able to prepare large jobs.

A cluster based on Intel EM64T processors and InfiniBand interconnect is added to the workbench. It serves both for post-processing/visualization and as a compute server for multi-disciplinary applications. The latter often require different types of architectures for different types of disciplines. The 200 node cluster is connected by a Voltaire InfiniBand switch with a bandwidth of 10 Gbit/s.

6.2 Fenfloss

The Institute for Hydraulic Machinery (IHS) of the University of Stuttgart and HLRS commonly work on the Finite Element Numerical FLOw Simulation System (Fenfloss), an integrated environment to design and shape hydraulic turbines. Computer tools will be used to perform a design for each specific water power plant. Numerical simulations warrant a high quality of the final design and optimize the overall efficiency. Insights gained during the analysis of flow simulations will thus immediately lead to design modifications.

The simulation process chain is partially established in the integrated environment. It is based on a parameterized runner design that enables a numerical optimization of axial and radial hydraulic turbines. COVISE, a distributed Collaborative Visualization and Simulation Environment developed by HLRS [8], is used as integration platform for the profile generation, the runner contour generation and the grid generation of the entire machine. The definition of boundary conditions based on the operating point, the coupled simulation around runner and guide wheel and the overall process chain will be controlled from within a virtual reality environment.

Fenfloss exhibits strong scaling with the number of nodes and has been found to reach a 50 percent efficiency on the NEC SX-8.

6.3 Demonstration in the HLRS Cave

Scientific visualization techniques are used by scientists and engineers to understand complex simulations [8]. They comprise filtering data, generation of deduced information, mapping of data onto visual representations and their display, finally. Distributed software environments often couple simulations on remote machines with local visualizations.

Virtual Reality techniques (VR) complement visualization methods to improve the comprehension of complex content and spatial relationships. Stereo projection rooms are used by groups of experts to enter a three dimensional virtual world consisting of visualized data or geometric representations of engineering parts. In such environments users are able to perceive interrelationships of complex structures and navigate with them. Interactions such as inserting particles into a flow field become possible, as can be seen in an MPEG movie of a virtual reality demonstration of a complete water power plant at Kiebingen near Stuttgart.

6.4 Virtual Tour of Kiebingen Water Power Plant

A Covise (Collaborative VISualization Environment) is used to demonstrate the water flow through a parameterized radial water turbine by means of simulation steering.

The animation shows: water flow lines with velocity along lines ranging from blue (slow) to red (fast). The field in the ISO cutting plane represents the radial component of water velocity (i.e. the angular momentum being applied on the runner wheel).

In a short pause of the simulation, the inclination of the blades in the guide wheel will be modified. The simulation is being restarted leading to the following steps: generation of a new mesh, decomposition of the compute problem and assignment to a number of parallel processors. After a delay of $\approx 10\,\text{s}$, first results are output by the visualization pipeline.

The radial turbine is completely parameterized: e.g. concerning the blade profile and geometry, number of blades, number of input/output water channels, etc.

It is important to notice that we are looking at an online simulation here which allows the control of most of the important design parameters by simulation steering, thus leading to a new, inductive way of design — even in very complex environments.

7 Conclusion

The HLRS Teraflop Workbench Project has resulted in a robust, scalable high performance computing environment that allows for the seamless integration of new systems and software over time. Applications from the engineering sciences like computational fluid dynamics, combustion, structural mechanics, and process engineering have been found to provide a good, if not excellent match to the available architectures. New users and new application fields have been brought to the supercomputer already. Also, research in storage and communication systems has enabled a new I/O culture that will enable a geographically distributed version of the HLRS Teraflop Workbench in the near future.

Acknowledgements We wish to thank the many developers within the Teraflop Workbench Collaboration who have created this unique platform and provided helpful comments on this paper. This work was in part performed by Alcatel SEL, Deutsches Zentrum für Luft- und Raumfahrt, European HPC Technology Centre, Nokia Siemens, Institute of Aerodynamics and Gasdynamics, Institute for Hydraulic Machinery, Institute of Process Engineering and Power Plant Technology, Institute of Structural Mechanics (all institutes at the University of Stuttgart), NEC, and finally the HLRS team. Special thanks are due to the HLRS Visualization Department for providing the virtual reality animations.

References

1. IEEE Datacenter Ethernet Call for Interest (CFI). http://www.ieee802.org/3/cfi/0304_1
2. IEEE 802.3ar Congestion Management Working Group. http://www.ieee802.org/3/ar
3. HPSS System Administration Guide: High Performance Storage System Release 4.1., IBM (1998)
4. 6th HLRS/hww Workshop on Scalable Global Parallel File Systems, April 16–18, HLRS, Stuttgart (2007), http://java.hlrs.de/corga-hwws-2007
5. Teraflop Workbench Initiative. http://www.teraflop-workbench.de
6. Resch, M., Küster, U., Müller, M., Lang, U.: A Workbench for Teraflop Supercomputing. In: SNA03, September 22–24, Paris (2003)
7. Watson, R.: High Performance Storage System Scalability: Architecture Implementation and Experience. In: Proceedings of the 22nd IEEE / 13th NASA Goddard Conference on Mass Storage Systems and Technologies, April 11–14, pp. 145–159, Monterey (2005), http://www.hpss-collaboration.org/hpss/about/watsonr_highperf.pdf
8. Wössner, U.: Scientific Visualization and Virtual Reality, High-Performance Computing and Communication, HLRS, Stuttgart (2005)

Development of Cognitive Methodology based Data Analysis System

Chiaki Kino, Yoshio Suzuki, Noriyuki Kushida, Akemi Nishida, Sachiko Hayashi, and Norihiro Nakajima

Abstract Nuclear engineering is an integrated engineering field of mechanical and civil engineering, partical physics as well as fluid and thermodynamics. Researchers in nuclear engineering fields need to treat extensive physical and engineering information obtained through theories, simulations, experiments and observations in order to promote a nuclear technology safely and securely. To meet the need, the Cognitive methodology-based Data Analysis System (CDAS) which equips information technologies that have recognition abilities similar to those of humans has been developed. The system supports researchers to analyze numerical simulation data by using extensive scientific knowledge. In the present study, information technology is developed for performing these processes and for configuring systems. In addition, a prototype system has been constructed using this information technology and an application experiment using a virtual plant vibration simulator has been performed to confirm the implementability of the system. The results obtained demonstrate that the CDAS enables researchers to dynamically set essential functions for evaluation and judgment, enabling them to readily extract meaningful and reliable information from large-scale data of up to 1 TB.

1 Introduction

Having an accurate knowledge of the physical phenomena related to nuclear technology is critical for ensuring the safe usage of nuclear technology. To achieve this, it is essential to consider all possible physical and engineering data obtained through theory, simulations, experiments and observations. Researchers need to an-

---

Chiaki Kino · Yoshio Suzuki · Noriyuki Kushida · Akemi Nishida · Sachiko Hayashi · Norihiro Nakajima
Japan Atomic Energy Agency, 6-9-3 Higashi-Ueno, Taito-Ku, Tokyo 110-0015, Japan, e-mail: {kino.chiaki, suzuki.yoshio, kushida.noriyuki, nishida.akemi, hayashi.sachiko, nakajima.norihiro}@jaea.go.jp

alyze ever-increasing amounts of data as theories become more sophisticated, computer and network techniques become more advanced and the performance of measurement systems increases. Recently, computational science and engineering have become established methodologies for conducting research and design, and they supplement the traditional theoretical and experimental approaches. Nuclear engineering simulations are generating much more data than experiments because of their spatial and temporal scalability and their ability to model a wide range of phenomena. As a typical example, a seismic response analysis of an entire nuclear power plant has been conducted [7] to model the physical phenomena that occur in a nuclear power plant operating during an earthquake. Real scale experiments on a shaking table are highly impractical due to the huge costs and long periods involved. Thus, numerical simulations play an important role in analyzing the effect of an earthquake on the operation of a nuclear power plant.

Data analysis of numerical simulations involves extracting quantitative and/or qualitative information from the data generated. When the result data includes a huge amount of data relating to specific components, models, materials and engineering specifications, it is difficult for researchers to judge whether the data is sufficiently accurate, or whether the data contains information about potentially serious problems such as stress concentrations and cracks. In the case of a seismic response analysis, an entire nuclear power plant consists of over 2000 parts and each individual part has its own properties, material models, and engineering specifications. Consequently, the amount of data generated is of the order of 1 TB. Although researchers evaluate the result data by using appropriate evaluation indices and judgment criteria, there is concern that human errors, such as oversight or misunderstandings, are increasing due to the proliferation in the number of indices and criteria. Researchers therefore require an efficient system that assists them to process large and complex sets of data and that minimizes the possibility of oversights or misunderstandings.

Some data analysis systems specifically designed for nuclear engineering are being developed in order to meet these requirements [11, 6, 13, 5, 8]. Shu [11] proposed an information inference system which is based on an artificial neural network method to evaluate the structural health status of a nuclear power plant. Another example is the expert system proposed by Laraki et al. in [6] for the non-destructive control of a plant by using a gamma-ray scanning technique. Conventional data analysis systems adopt procedural methods based on the assumption that these systems target a specific problem. Generally, data analysis for numerical simulation targets various problems depending on the researchers purpose or the objectives of the simulation. It is difficult for conventional data analysis systems that are based on such an assumption to conduct data analysis of numerical simulations.

Our center has been developing the Cognitive methodology-based Data Analysis System (CDAS), which enables information technologies that have recognition abilities similar to those of humans [4]. This system aims to reduce the burden on the researcher when analyzing large-scale data sets, thus reducing the possibility of human errors. To realize the CDAS, recognition processes common to all data analysis systems have been adopted and usable informational technologies have been proposed that have recognition abilities without targeting a specific problem. The focus of the present study is on evaluation and judgment processes in order to reduce the burden on researchers. Additionally, a system configuration for executing these technologies has been designed in the present study.

The present paper discusses whether the proposed technologies and system configuration are effective in reducing the burden on the researcher. To achieve this, a prototype system having the above-mentioned information technologies has been constructed. This system has been implemented in a grid environment in order to overcome the difficulty of processing large-scale data. Moreover, the stress distribution of a virtual plant vibration simulator has been analyzed as an application experiment.

2 The Basic Idea of CDAS

We first discuss the data analysis process and the data evaluation and judgment structure. In addition, we clarify the necessary functions and abilities for data analysis.

2.1 Data Analysis Process

First, it is necessary to select the required functions to perform data analysis of numerical simulations. We thus consider numerical simulation and data analysis in some depth. In the case of numerical simulations, researchers need to construct a model, formulate the physical phenomena of the system and define the design data that includes the kind of components and materials of the system. Next, is necessary to determine numerical solutions and define analysis data that includes the mesh size and boundary conditions. Subsequently, a numerical simulation is used to solve the equations by employing numerical solutions and analysis data. After performing the calculations, the numerical simulation outputs the result data. Researchers analyze the result data and draw conclusions. The same process is equally valid in an experimental or theoretical study. In the case of an experimental study, researchers need make a model and construct experimental equipment for the model. They can then observe the phenomena and acquire data by using the experimental equipment.

2.2 VV and DD Functions

In order to arrive at conclusions, it is necessary to evaluate the experiments reliability by Validation and Verification (VV) [9] since modeling and solutions always include certain assumptions and approximations. Here, validation means solving the correct equations while verification means solving the equations correctly. Essentially, researchers validate the physical assumptions or the hypothesis of a model and verify the approximation error of numerical solutions or the truncation error of a numerical simulator. Additionally, result data contains some scientific knowledge, so researchers must discover the scientific characteristics from reliable data (Data Diagnosis: DD). Finally, they attempt to draw conclusions by considering validated and meaningful information. The CDAS requires both VV and DD functions in order to arrive at conclusions.

2.3 Synthesis Function

Moreover, the CDAS must process comprehensively all the result data from a numerical simulation and output the data from the VV and DD functions. Thus, the CDAS requires a synthesis function to integrate all the information. This function has the ability to recognize the required information and the ability to interpret the findings from the VV and DD functions.

2.4 Evaluation and Judgment

The purpose of analysis is to acquire information for evaluating and judging the validity and meaningfulness of the data: evaluation index and judgment criteria are used to achieve this. Evaluation involves extracting the evaluation index of the analysis target by using engineering information. Judgment involves evaluating the meaningfulness of the analysis target by using the evaluation index and judgment criteria.

Judgment is categorized into two types: quantitative judgment and qualitative judgment. Quantitative judgment is based on deviation from a criterion. For example, we can adopt the stress distribution of a standpipe as an analysis target, the maximum value of the stress distribution as an evaluation index and the yield stress as a judgment criterion. If the maximum value of the stress distribution is greater than the yield stress, the stress distribution can be judged to be dangerous. On the other hand, a qualitative judgment uses pattern recognition. For example, the stress distribution in a standpipe can be taken as an analysis target, the shape of the stress distribution as an evaluation index and a convex upward stress distribution as the judgment criterion. If the shape of the stress distribution is convex upward, it can be judged that a stress concentration exists in the region. In the present study, these structural analysis evaluations are implemented using the finite element method (FEM).

3 System Configuration

Figure 1 shows a system configuration of the CDAS. The CDAS consists of three functions, namely a synthesis function, a verification and validation (VV) function, and a data diagnosis (DD) function. The VV and DD functions consist of data evaluation and judgment abilities. Synthesis functions consist of cognizance ability of engineering information and targets, indices, criteria and interpretation ability of the VV and DD functions. The VV and DD functions are distributed on a grid in order to process large-scale data sets. The details of these functions are discussed in more depth later.

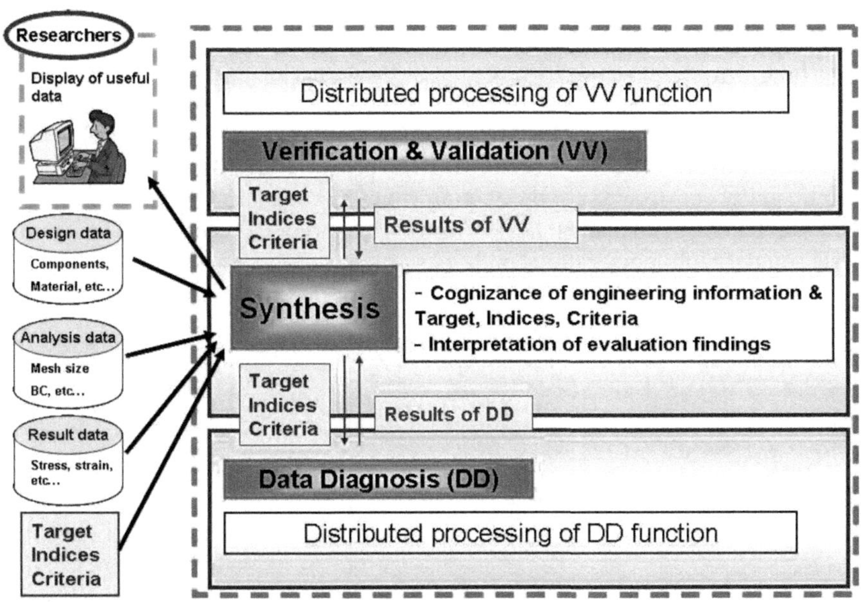

Fig. 1 System configuration of CDAS

3.1 *Flow of Data Analysis by CDAS*

The CDAS requires design data, analysis data and result data as input data. Additionally, the VV and DD functions require targets, indices and criteria. Firstly, the synthesis function cognizes the information. Secondly, the CDAS selects the necessary targets, indices and criteria for the VV and DD functions and sends them to these functions. Next, the VV and DD functions are carried out separately and return the results to the synthesis function. The synthesis function interprets the results and displays only the useful information to researcher.

3.2 Implementation on a Grid Computing

The CDAS has various methods for evaluation and judgment. For efficient computation, these methods are executed on supercomputers in a grid computing environment. The grid middleware used in this study was the Information Technology Based Laboratory (ITBL) and Atomic Energy Grid Infra Structure (AEGIS), which is a specialized structure developed by our center especially for nuclear engineering and science [12].

4 Result and Discussion

As the first step of our study, we restricted the target of the CDAS to the large amounts of data obtained from a large-scale structural analysis using the FEM [15]. The system was applied to the virtual plant vibration simulator of the High-Temperature Engineering Test Reactor (HTTR) [1] shown in Fig. 2. The HTTR is a research facility constructed for developing High-Temperature Gas-Cooled Reactor (HTGR) technology and nuclear heat utilization technology as well as for irradiating materials under high-temperature conditions.

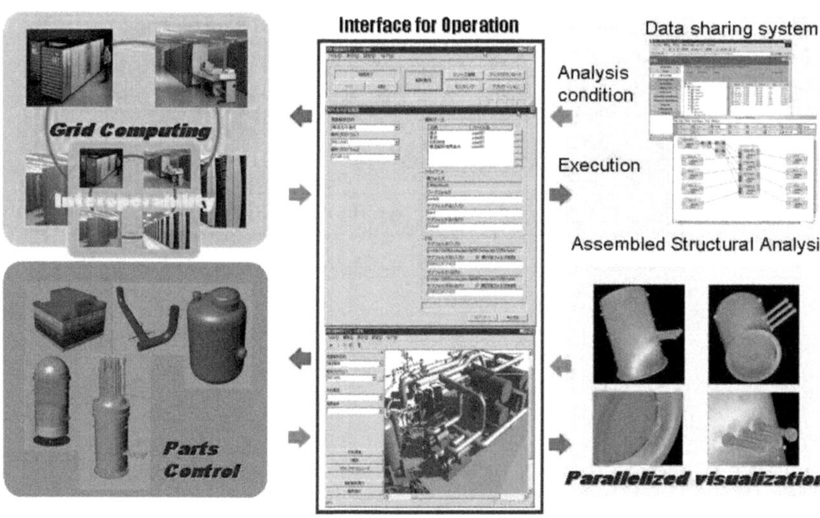

Fig. 2 Outline of three-dimensional virtual plant vibration simulator

The objective of this simulator is to reveal the seismic response of an entire nuclear power plant. Even for a simple model, the simulator has over 7.5 million nodes, over 23 million degrees of freedom and a dataset of over 1 TB. Thus, the CDAS plays an important role in this large data analysis. In the present study, the virtual nuclear power plant is loaded 6.0×10^4 (N) in the horizontal direction. This value was calculated from Japanese earthquake-resistance standards. A real scale experiment on a shaking table is impossible due to the huge cost and long time required. It is thus very important to thoroughly examine stress distributions under such conditions.

Estimation of the mesh dependency by the ZZ method [16] is used as an example of an implementation of verification. This method is a kind of quantitative judgment. It uses the error energy norm as an index. The error energy norm is obtained from the difference between the smoothing value and the numerical solution, and when this value is below the criterion value, the numerical solution is judged to be convergent.

However, there many other methods for obtaining the energy norm. Each method has its own properties. It is thus not simple to determine which method is the most appropriate. Similarly, there are a lot of criteria for determining convergence. Again, it is not simple to judge which values are the most appropriate. So, we need to combine various methods. In this study, the CDAS evaluates the error energy norm by combining the Winslow method [14] and SPR [2].

In this study, CDAS used quantitative judgment using yield stress which is obtained from material information with qualitative judgment using Neural Networks. When the maximum value is over this criterion value and the shape of the distribution is a quadratic curve which is convex upward, this function judges the stress distributions as dangerous stress concentration.

On the other hand, we adopted a neural network [10] using back propagation in order to evaluate and judge whether the stress distributions are meaningful or not. A neural network using back propagation with a single hidden layer has been demonstrated to approximate arbitrary continuous functions [3]. The stress concentration can be defined in terms of stress distributions with peaks. That indicates a convex-upward quadratic curve. Therefore, a neural network is used to find such a quadratic curve. This method is a kind of qualitative judgment.

Figure 3 shows a GUI of the current CDAS for structural analysis. The GUI has the following functions:

1. Selection of a data file to be read and display a mode,
2. Selection of a target to analyze
3. Determine a region to evaluate and judge data
4. Run a scan
5. Display of results (see Fig. 3)

Figure 3 also shows the stress distributions suggested by the CDAS. The CDAS located the peak stress at the base of the standpipes. In addition, the validity of the data in the region is confirmed by the ZZ method. Consequently, CDAS evaluated and judged that the peak is valid and meaningful from a engineering viewpoint and it displayed the data. Up until now, it has been impossible to analyze an entire nuclear power plant due to the large amount of data involved. However, by applying the CDAS to this analysis, we succeeded in thoroughly analyzing this data for the first time.

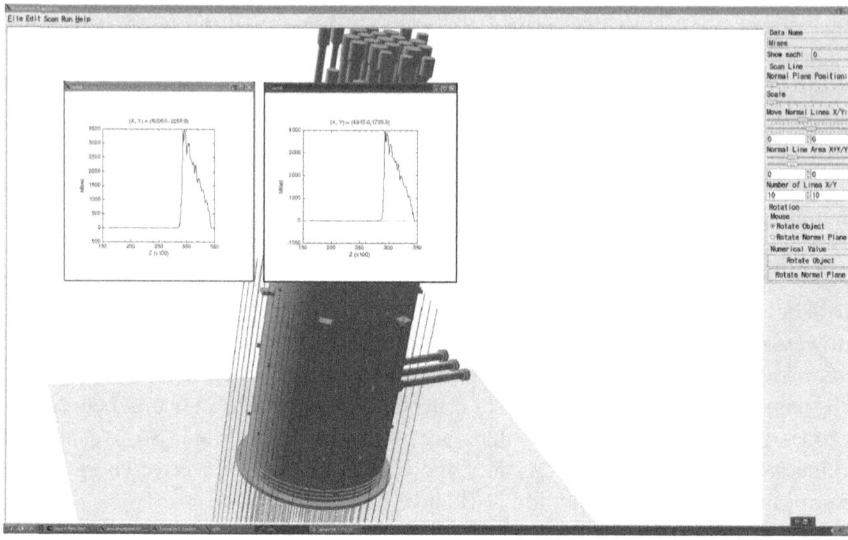

Fig. 3 Display of results by CDAS

5 Conclusions and Future Works

The basic concept of the CDAS has been proposed. This system enables researchers to analyze all possible data from nuclear engineering fields. To realize this system, information technology for substituting the human recognition abilities and a system configuration has been designed. It is shown that the synthesis function is very important to process the result data from a numerical simulation and output the data from the VV and DD functions. Additionally, it is shown that recognition processes of evaluation and judgment are describable by analysis targets, evaluation index and judgment criteria.

In the present study, a prototype system with information technologies has been constructed on a grid environment and applied to structural analysis in order to confirm the implementability of the system. The present results demonstrate that the system enables a reliable and meaningful stress concentration to be determined from a large-scale data set of up to 1 TB without human assistance.

The CDAS can dynamically set arbitrary analysis targets, evaluation indices and judgment criteria and enables the thorough evaluation and judgment of large-scale data by using all possible information without relying on human assistance. At this point, we can say that the CDAS reduces the researcher's burden. It is anticipated that this system will be applicable to other scientific fields. In the future, the amount of data will be increased, so that the CDAS will be more useful, and even become an essential system.

Acknowledgements We would like to thank JAXA (Japan Aerospace Exploration Agency), JAIST (Japan Advanced Institute of Science and Technology) and HLRS (High Performance Computing Center Stuttgart) for usage of their computing system under ITBL and AEGIS.

References

1. HTTR, http://www2.tokai-sc.jaea.go.jp/httr/eng/index_top_eng.html
2. Ainsworth M., Oden J. T.: A Posteriori Error Estimation in Finite Element Analysis, Wiley Inter-Science (2000)
3. Cybenko G.: Approximation by superpositions of a sigmoidal function, Math. Control Signals Systems, **2**, 303–314 (1989)
4. Kino C. et al.: Cognitive methodology based Data Analysis System for Large Scale Data, SC07 HPC Analytic Challenge (2007)
5. Kobare S. K.: Expert systems for emergency alarms analysis during accident situations in nuclear reactors, Reliab. Eng. Syst. Saf., **37**(2), 139–149 (1992)
6. Laraki K., Alami R., El Moursli R. Cherkaoui, Bensitel A., El Badri L.: An expert system for improving the gamma-ray scanning technique, Nuclear Instruments and Methods in Physics Research Section A: Accelerators, Spectrometers, Detectors and Associated Equipment, **578**(1), 340–344 (2007)
7. Nakajima N., Matsubara H., Minami T., et al.: The Large Scale Numerical Analysis of Nuclear Power Plant in Distributed Computational Environment, SC05 HPC Analytic Challenge (2005).
8. Ouyang J. et al.: Modeling of PWR Plant by Multilevel Flow Model and its Application in Fault Diagnosis, J. of Nuclear Science and Technology, **42**(8), 695–705 (2005)
9. Roache P. J.: Verification and Validation in Computational Science and Engineering, Hermosa (1998)
10. Rumelhart D. E., Hinton G. E., Williams, R. J. In: Parallel Distributed Processing, Explorations in the Microstructure of Cognition. Vol. 1: Foundations, pp. 318–362, MIT Press (1986)
11. Shu Y.: Inference of power plant quake-proof information based on interactive data mining approach, Advanced Engineering Informatics, **21**, 257–267 (2007)
12. Suzuki Y., et al: Research and Development of Fusion Grid Infrastructure Based on Atomic Energy Grid InfraStructure (AEGIS), Fusion Engineering and Design, **83**, 511–515 (2008)
13. Takiguchi H., Takamatsu H., Uchida S., Ishigure K., Nakagami M., Matsui M.: Water Chemistry Data Acquisition, Processing, Evaluation and Diagnostic Systems in Light Water Reactors, J. of Nuclear Science and Technology, **41**(2), 214–225 (2004)
14. Winslow A. M.: Numerical Solution of the Quasi-linear Poisson Equation in a Nonuniform Triangle Mesh, J. Comput. Phys., **2**, 149–172 (1967)
15. Zienkiewicz O. C., Taylor, R. L.: The Finite Element Method (Fifth edition), Betterworth–Heinemann (2000)
16. Zienkiewicz O. C., Zhu J. Z., Int. J. Numer. Methods Eng., **24**, 337–357 (1987)

IV
Chemical Applications

3D-Flame Modelling in Power Plant Applications

Benedetto Risio, Norbert Paßmann, Friedhelm Wessel, Egbert Reinartz

Abstract Simulation based optimisation of boiler operation and boiler design is a key technology for minimizing investment and operation costs of coal fired power plants. Simulation based optimisation requires to transform the real plant process into the virtual reality of a mathematical 3D-boiler model. Potential combustion problems (e.g. wall corrosion, increased carbon loss, etc.) can be identified in the virtual reality and be avoided before they become expensive reality. Today's boiler models range from 5-10 Mio. grid points of resolution. Such high resolution models have to be calculated on high-performance supercomputers like the high performance computing installation NEC-SX8. This paper is intended to present the current status of boiler modelling technology.

1 Introduction

In the past, the design of large scale combustion equipment was mainly based on empirical and experimental methods. The technical risks involved with the investment decisions that targeted design changes were therefore significantly high. Experience gained from the use of simulation models during recent years show that the knowledge gained by 3D flame modelling permits a reliable evaluation of boiler reconstruction measures today. Potential problems are identified early in advance

Benedetto Risio
RECOM Services GmbH, Nobelstraße 15, 70569 Stuttgart, Germany, e-mail b.risio@recom-services.de

Norbert Paßmann · Friedhelm Wessel
Kraftwerk Neurath, RWE Power AG, Energiestraße, 41517 Grevenbroich, Germany, e-mail: {norbert.passmann, friedhelm.wessel}@rwe.com

Egbert Reinartz
RWE Power International RE GmbH, Stüttgenweg 2, 50935 Köln, Germany, e-mail: egbert.reinartz@rwe.com

and reliably avoided before these problems become expensive reality. The methodology of applying 3D-flame modelling is exemplified at a burner retrofit project in the brown coal fired power plant Neurath.

The Neurath power plant, consisting of 5 units (3×317 MWel and 2×634 MWel) with a total 2,219 MWel is used in base-load operation. In the late 80s, the furnaces were modified to low-NOx combustion. This helped reduce NOx emissions to values $< 200\text{mg}/\text{m}^3$. In line with structural conditions, overfire air was injected at several levels and cold gas recirculation provided above the burner belt, among others. In 2000, furnace reconstruction measures were initiated due to changed brown coal qualities and resulting operational restrictions in units A and B of the Neurath power plant. Among others, the existing multi-jet burners were replaced by swirl burners made by Babcock Hitachi Europe. To ensure that the actual design corresponds to the target design and allow further optimisation of the furnace once the reconstruction is completed, RECOM Services GmbH was awarded a contract to perform 3D computer simulations of the combustion performance and fouling behaviour.

2 Flame modelling tool and computer hardware

The simulation tool used in the present work is the 3D flame modelling programme RECOM-AIOLOS. In-depth validation using measuring data from industrial power plants [2, 3, 4], the extension of chemical reaction models [1] and rapid development of computer technology have made RECOM-AIOLOS a well proven and reliable tool for the prediction of industrial furnace efficiency. RECOM-AIOLOS includes among others models for the evaluation of wall corrosion risk, burnout, emissions or fouling, so that 3D flame modelling can make reliable predictions for a large number of operation-relevant issues. To this end, the actual firing process in the power plant is mapped in the virtual reality of a mathematical 3D furnace model. The physical furnace space is approximated with a multitude of so-called finite volumes. Today, RECOM typically uses up to 10 million volumes for geometrical mapping of a furnace. For each of these volumes, the RECOM-AIOLOS flame modelling programme forms conservation equations for mass, momentum, energy and element conservation. All in all, some 100 equations are formed for each volume. To visualize an operation mode in the virtual reality, RECOM-AIOLOS must find a solution for the coupled equation system of 100×10 million $= 1$ billion equations. The resulting computational demands can only be met by particularly high-performance computers as those used in climate research centres for weather forecasting. The High Performance Computing Center Stuttgart (HLRS) has such a powerful system.

The high performance supercomputer NEC-SX8 (see Fig. 1) installed at HLRS has a 9.2 TB main memory and 72 nodes, each equipped with 8 high performance vector processors and reaches a total peak performance of 12 trillion floating-point operations per second. This capacity is sufficient to solve the required simulation problem using up to four nodes (equivalent to 32 processors) of the overall system

Fig. 1 Vector supercomputer NEC-SX8 at the High Performance Computing Center Stuttgart (HLRS)

for a maximum of 12 hours. By contrast, the solution of this problem on standard PC hardware would take several weeks. For viewing and analysis of the simulation results conventional screen-based methods and increasingly real 3D visualization methods (virtual reality) are being used. Here, in a room, the impression can be generated that the viewer is actually in the virtual furnace as shown in Fig. 2. He can view the simulated results in a three-dimensional manner and also move freely in the virtual furnace.

Fig. 2 RECOM employee analysing the flame structure of a burner in the virtual reality environment at HLRS

3 Neurath A/B boiler model

The furnace model of the Neurath units A and B prepared in 2000 with only some 1.5 million grid points has been continuously improved and developed further over the years. The 18 swirl burners and the related burner nearfields are represented today by a total of some 5.41 million grid points. This corresponds to a resolution of around 1–10 cm in all directions in space (see Fig. 3). The remaining furnace is mapped with a resolution of approx. 2–20 cm and some 4.16 million grid points. The resolution of the convective part is about 16–30 cm in all directions in space (total of some 310,000 grid points). Figure 4 shows the current overall boiler model with some 9.9 million grid points and integrated mapping of the convective part.

Besides boiler geometry, the operating parameters of the power plant must also be defined in the model. Extensive mill measurements of the dust and carrier gas distribution as well as I&C data are used to determine:

- the distribution of brown coal and air mass flows to the various burners and overfire air nozzles,
- the temperatures of carrier gas and secondary air,
- the recirculated flue gas quantity, final water content of coal in the carrier gas, coal composition (ultimate analysis) and net calorific value,
- coal fineness (particle size) of mills,
- the swirl number and sense of rotation of the burners and
- the steam-side parameters (flow rates, temperatures and pressures) for the evaporator and convective heat exchangers.

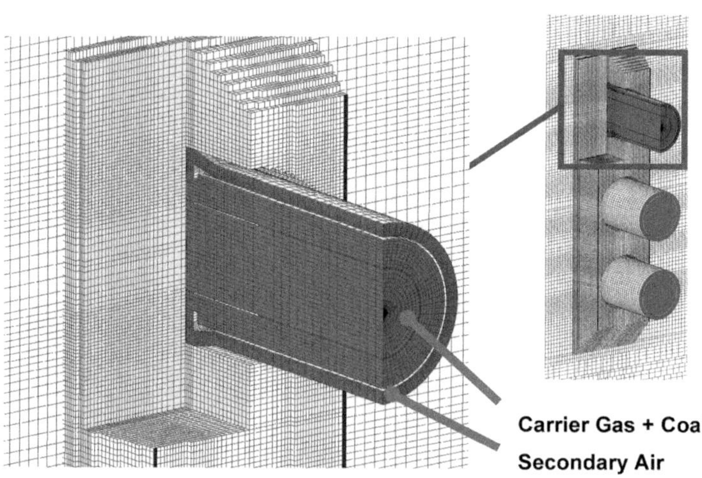

Fig. 3 Burner and burner nearfield in a 3D resolution of some 1–10 cm in all directions in space

4 Simulation results

Results obtained by the computer simulation include both three-dimensional distributions of flow velocity, temperature, species concentrations (e.g. O_2, CO, NOx, unburnt carbon) and two-dimensional distributions of wall atmosphere, wall temperature, wall heat fluxes, particle impact and deposition rates. To visualize these detailed results, 3D and 2D plots, isosurfaces and x-y diagrams can be extracted from the total dataset. The high spatial resolution of the boiler model allows both realistic representation of the processes in the burner nearfield and the interaction of the burners in the overall firing system.

As examples, Figs. 5–7 show velocity, CO and temperature distribution in the burner nearfield of Mill 3 (bottom level). The figures illustrate the complexity of the flow processes and the resulting reaction processes as well as the necessity of high resolution selected for the current model.

Since, due to the uneven dust and carrier gas distribution and the wear conditions of the various mills, the operating parameters of each burner are differing, each mill combination results in different burner interaction in the overall firing system.

Figure 8 illustrates this fact, taking the example of temperature distribution on the bottom burner level for the simulated operation mode of October 1, 2004, with Mills 4 and 6 out of operation.

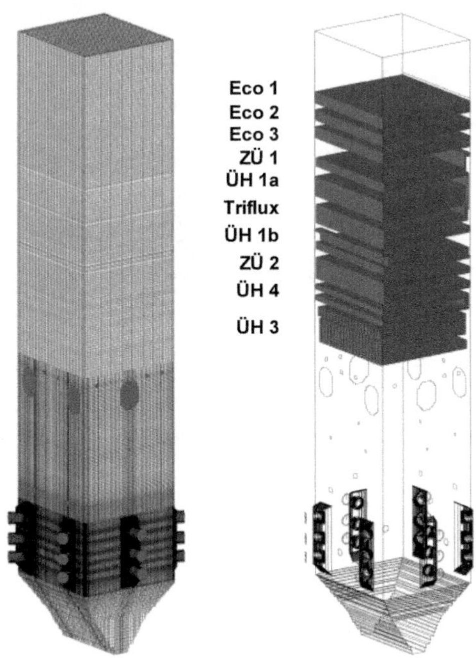

Fig. 4 Present overall boiler model of Neurath units A/B with 9.9 million cells and integrated mapping of the convective part

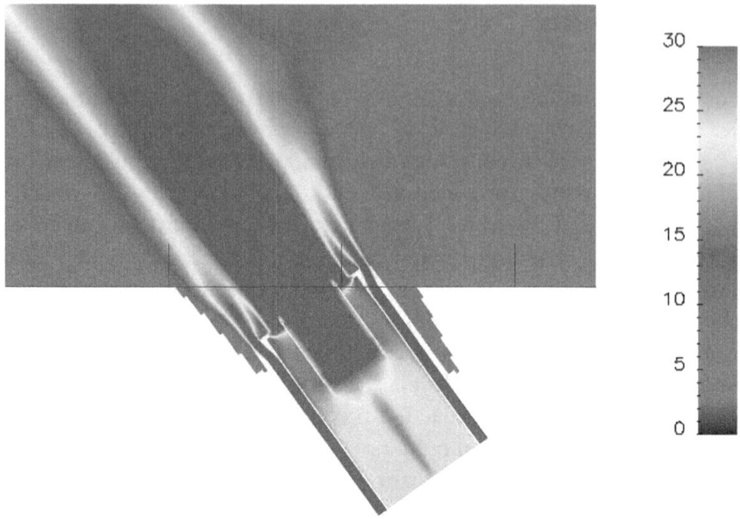

Fig. 5 Velocity distribution (magnitude) [m/s] in the burner nearfield

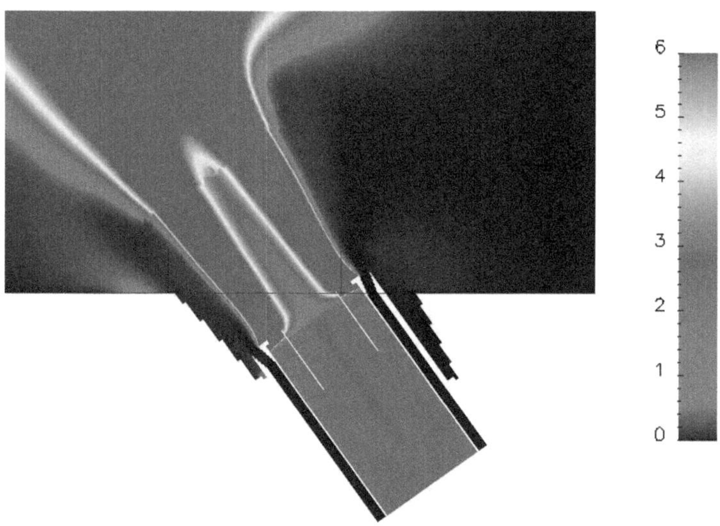

Fig. 6 CO distribution [% vol., dry] in the burner nearfield

The fire location resulting from interaction of all burners in the firing system is visualized in Fig. 9 by showing the temperature distribution as an isometric projection with a 1,300°C isosurface (grey). All temperatures inside the surface are above 1,300°C and all temperatures outside the surface below 1,300°C. Figure 9 clearly shows the shift of the fire towards the front wall and the left sidewall, which is the result of the selected mill combination and burner interaction in the firing system. Detailed knowledge of the fire location and hence local temperatures and

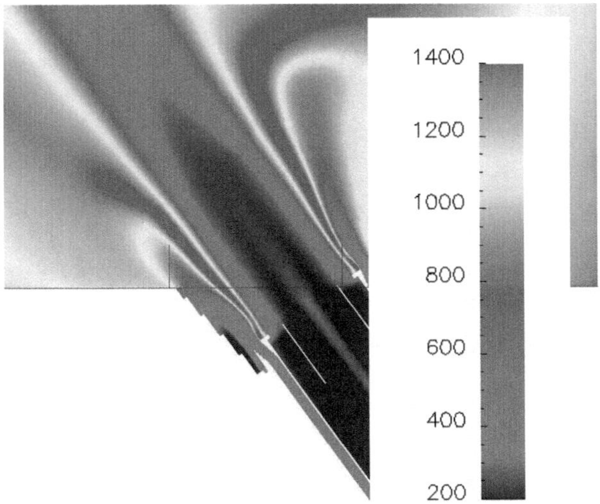

Fig. 7 Temperature distribution [°C] in the burner nearfield

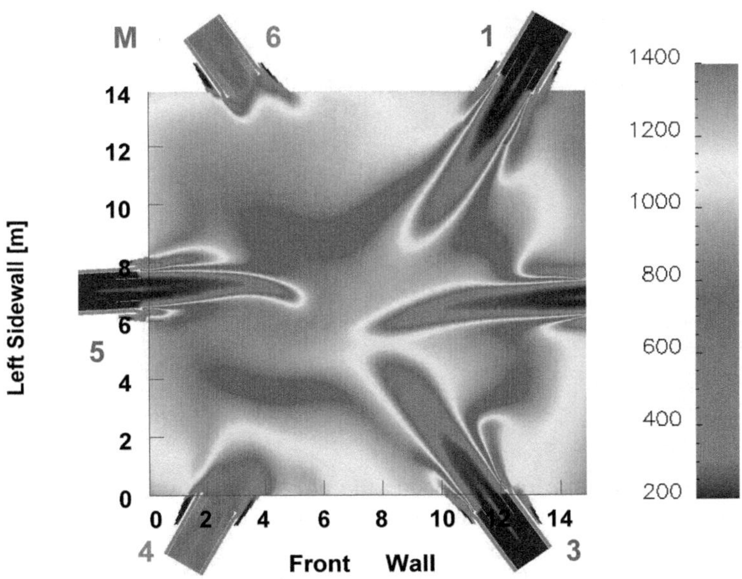

Fig. 8 Burner interaction illustrated by means of the temperature distribution [°C] on the bottom burner level

stoichiometries allows an analysis of the NOx formation processes and fuel burnout in the firing system.

Figures 10 and 11 show the pattern of NO formation and CO burnout plotted against boiler height. The patterns show the average NO and CO concentration for the respective elevation. They reflect the significantly reduced NO production that

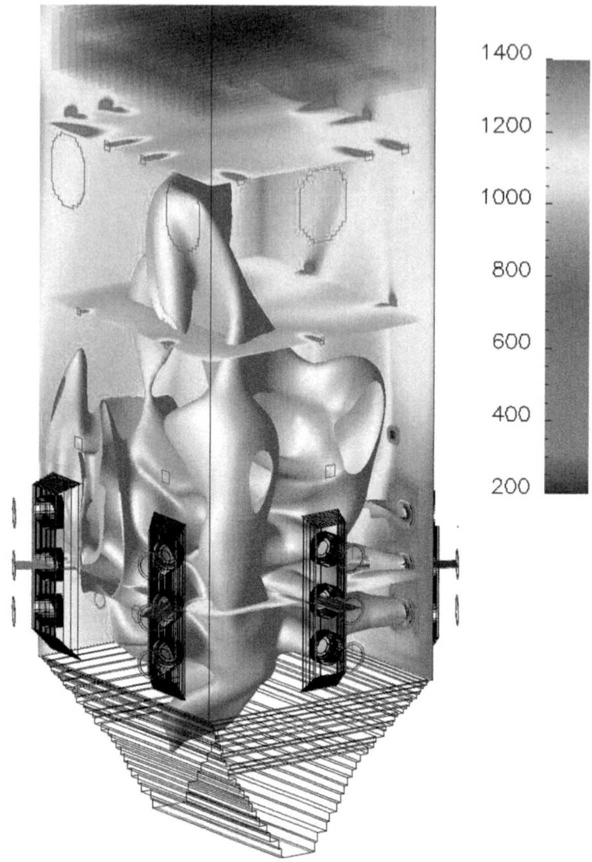

Fig. 9 Temperature distribution [°C] as 3D projection with 1,300°C isosurface (grey) to visualize the fire location

results from temperature reduction due to cold gas injection. For overfire air injection, the first effect is dilution followed by production induced by the burnout process accelerated by oxygen addition (see Fig. 11).

The basic prerequisite for a successful use of computer simulation to answer operational questions is a continuous comparison of simulated values with measured values from power plant operation. Comparative values from power plant operation are e.g. CO and NOx concentration at the furnace exit. Table 1 shows a comparison of calculated averages with operational measurements for a boiler measurement of November 28, 1998 with conventional multi-jet burners and blended brown coal being used and a boiler measurement of October 1, 2004 with swirl burners and Hambach brown coal.

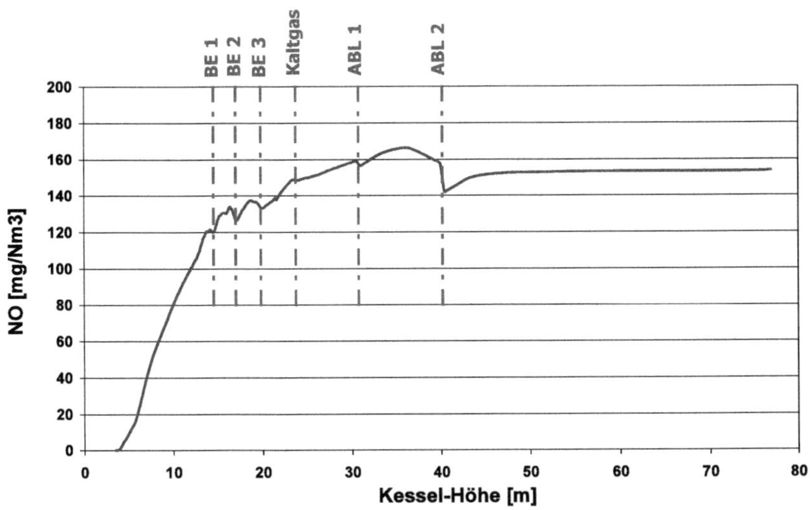

Fig. 10 NO formation plotted against boiler height as a consequence of cold gas and overfire air addition

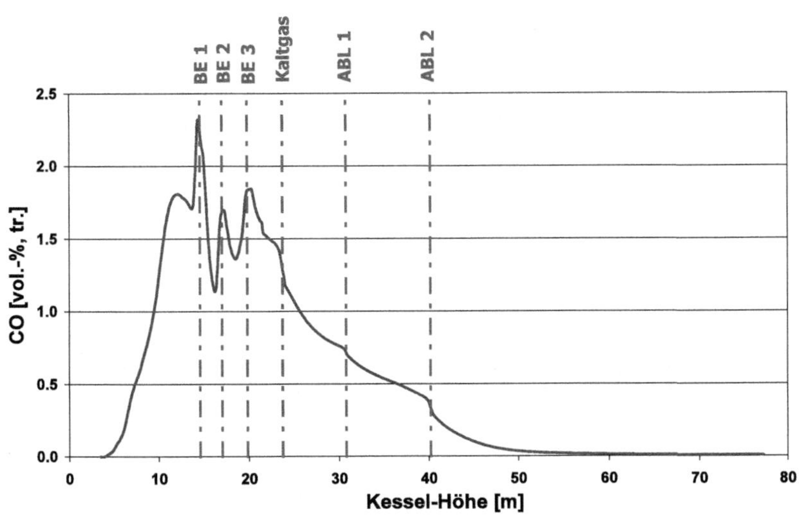

Fig. 11 CO burnout plotted against boiler height as a consequence of cold gas and overfire air addition

For swirl burners, the comparison in Table 1 indicates comparable NOx concentrations at the furnace outlet despite a significantly increased air factor in the burner belt. At the same time, a somewhat lower CO value emerges. The simulated values show sufficient agreement with the measured variables within the limits of some $\pm 10\%$.

Table 1 Comparison of calculated averages with operational measurements

	Multi-jet burner		Swirl burner	
	11/28/1998		10/01/2004	
Fuel	Blended brown coal		Hambach brown coal	
Air factor at upper burner edge	0.97		1.007	
	Measurement	Simulation	Measurement	Simulation
NOx mg/m_n^3	195	196	191	205
CO mg/m_n^3	89	98	75	84

5 Summary

This article reveals that current boiler models of RECOM Services with some 10 million grid points allow both a detailed visualization of the processes in the burner nearfield and interaction of the burners in the overall firing system. Even for such high-resolution models, the use of high-performance computers permits computing times of some 12h for individual calculation runs and makes these model sizes manageable. A comparison of calculated average values with operational measurements showed sufficient agreement of the calculated values with the measured variables within the limits of some $\pm 10\%$. When boiler reconstruction measures are evaluated, flame modelling gives detailed insight into the functioning of the planned measures. The knowledge gained from 3D flame modelling allows potential problems to be identified in advance, so that they can be reliably avoided when the planned measures are converted into reality.

References

1. Berreth, A., Han, X., Risio, B., Kein, K.R.G.: Einsatz detaillierter Gaschemie und nichtoxidierender heterogener Feststoffreaktionen zur Verbesserung der Verlässlichkeit von CO-Vorhersagen in CFD-Simulationen in Kohle- und Gasflammen. 22. Dt. Flammentag, VDI-Berichte, vol. 1888, pp. 471– 477, Braunschweig (2005)
2. Paßmann, N., Reinartz E., Tigges, K.D.: Feuerungsumbau und erste Betriebserfahrungen mit Rundstrahlbrennern im Braunkohlekraftwerk Neurath. VGB PowerTech, pp. 68-73 (2002)
3. Risio, B.: Effizienz und Verlässlichkeit numerischer Vorhersagen zur rechnergestützten Optimierung von Großkraftwerksfeuerungen. VDI Fortschritts-Berichte, Reihe 6 Energietechnik, vol. 471, Düsseldorf (2001)
4. Sabel, T., Käss, M., Kirschning, F.P., Greißl, O., Risio, B.: Betriebsoptimierung durch simulationsgestützten Feuerungsvergleich. 22. Dt. Flammentag, VDI-Berichte, vol. 1888, pp. 35-40, Braunschweig (2005)

Hierarchical Modeling of Combustion Processes

Ulrich Maas, Viatcheslav Bykov, Andriy Rybakov, and Rainer Stauch

Abstract Combustion processes are governed by a strong coupling of chemical kinetics, molecular transport processes and flow. Mathematical modeling is complicated by the existence of scaling problems (time-, velocity- and length scales). In order to allow a reliable numerical simulation of practical combustion systems, models have to be devised which do not neglect or over-simplify the underlying physical and chemical processes. In this paper hierarchical modeling concepts are presented which allow the development of realistic and reliable modeling tools based on information from detailed simulations.

1 Introduction

Reactive flows are governed by a complex interplay of chemical reaction, flow and molecular transport. They can be described mathematically based on conservation equations for mass, momentum, energy and species masses [13, 4, 35]. These conservation equations form a large system of stiff partial differential equations and therefore their solution is a great challenge. One major problem is that the description of chemically reacting systems leads very often to reaction mechanisms with far above hundred chemical species, i.e. over a hundred partial differential equations, which possibly react within more than a thousand of elementary reactions [35]. These kinetic processes cover time scales from nanoseconds to seconds. An analogous scaling problem arises for the length scales. The scales of technical systems may be several meters, and on the other hand, typical reaction zones or small turbulent structures have length scales of the order of millimeters. Due to these scaling problems the detailed simulation of three-dimensional turbulent flows in practical systems is beyond the capacity of even today's super-computers. Using simplified

Ulrich Maas · Viatcheslav Bykov · Andriy Rybakov · Rainer Stauch
Institut für Technische Thermodynamik, Universität Karlsruhe (TH), Kaiserstr. 12, 76128 Karlsruhe, Germany, e-mail: {maas, bykov, rybakov, stauch}@itt.uni-karlsruhe.de

sub-models is a way out of this problem. The question arising in mathematical modeling of reactive flows is then: How detailed, or down to which scale has each process to be resolved (chemical reaction, chemistry-turbulence-interaction, molecular transport processes) in order to allow a reliable description of the entire process. The aim is the development of models, which should be as simple as possible in the sense of an efficient description, and also as detailed as necessary in the sense of reliability. In particular, an oversimplification of the coupling processes between chemical reaction and turbulent flow should be avoided by all means to allow a predictive character. This is illustrated in Fig. 1, which shows the extreme range of spatial scales. In this work it is shown how hierarchical concepts can be used to solve this problem. Three examples are presented, namely a dimension reduction of the system of conservation equations based on the concept of intrinsic low-dimensional manifolds in composition space (ILDM) and reaction-diffusion manifolds (REDIM), a hierarchical description of the chemistry-turbulence interaction using a combination of direct numerical simulations and statistical models for the turbulent flow based on the solution of a transport equation for the joint probability density function of velocities and scalars, and a hierarchical description of multiphase processes based on detailed simulations of droplet combustion in combination with statistical models for the description of the spray. In these concepts information about processes on the smallest time and length scales enters the global models. Different examples show, that such hierarchical concepts allow a reliable description of reactive flows.

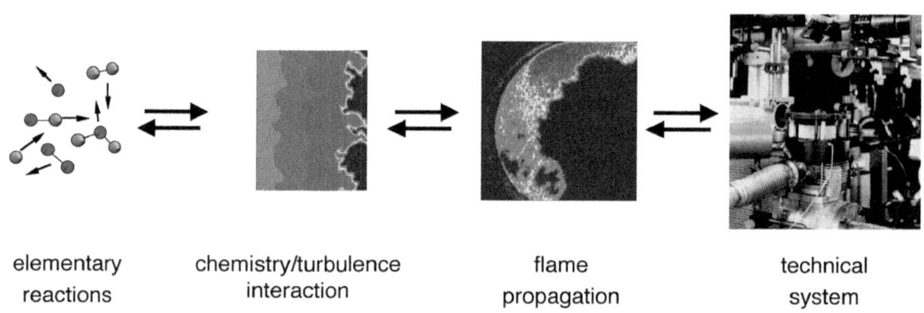

Fig. 1 Example for the different length scales in engine combustion. The mean free path of molecules is of the order of nanometers, whereas the characteristic flame thicknesses in the sub-millimeter range and the system dimension is of the order of several centimeters.

2 Dynamics of Reacting Flows

In its general formulation the dynamics of a reacting flow is governed by the set of conservation equations for mass, momentum, energy and species masses [13, 14].

$$\frac{\partial \rho}{\partial t} + \mathrm{div}(\rho \mathbf{v}) = 0 \qquad (1)$$

$$\frac{\partial \rho_i}{\partial t} + \mathrm{div}(\rho_i \mathbf{v}) + \mathrm{div}\mathbf{j}_i = M_i \dot{\omega}_i \qquad (2)$$

$$\frac{\partial (\rho \mathbf{v})}{\partial t} + \mathrm{div}(\rho \mathbf{v} \otimes \mathbf{v}) + \mathrm{div}\overline{\overline{p}} = 0 \qquad (3)$$

$$\frac{\partial \rho u}{\partial t} + \mathrm{div}(\rho u \mathbf{v} + \mathbf{j}_q) + \overline{\overline{p}} : \mathrm{grad}\mathbf{v} = 0 \qquad (4)$$

$$p = p(u, \rho_1, \ldots, \rho_{n_s}) \, , \qquad (5)$$

where t denotes the time, ρ the density, $\mathbf{v}$ the velocity, ρ_i the density of species i, $\mathbf{j}_i$ the diffusion flux density of species i, M_i the molar mass of species i, $\dot{\omega}_i$ the molar scale rate of formation of species i, $\overline{\overline{p}}$ the pressure tensor, u the specific inner energy, $\mathbf{j}_q$ the conductive heat flux density, $\overline{M}$ the mean molar mass and n_s the number of different chemical species. These equations hold for each phase, and the different phases are coupled via interface equations.

For the following analysis it is useful to separate the equations for the thermokinetic state variables from the equations for the flow field and to assume (for sake of simplicity) a low Mach number flow with constant thermodynamic pressure (a generalization to general flows is straight forward and shall not be considered here). The governing equation system for the scalar field can be re-written as

$$\frac{\partial \psi}{\partial t} = \mathbf{F} - \mathbf{v}\mathrm{grad}\,\psi - \frac{1}{\rho}\mathrm{div}\,(\mathbf{D}\mathrm{grad}\,\psi) \, , \qquad (6)$$

where $\psi = (h, p, w_1, \ldots, w_{n_s})^T$ (with h as the specific enthalpy, p the pressure, and w_i the species mass fractions (ρ_i/ρ)), $\mathbf{D}$ the matrix of transport coefficients (see, e.g., [11] for details), ψ the thermokinetic state vector (see above), and $\mathbf{F}$ the vector of source terms. The source terms $\mathbf{F}_i$ as well as the transport matrix $\mathbf{D}$ are complicated nonlinear functions of the thermokinetic state vector ψ [14, 11, 4]. Because (6) does not invoke any modeling procedure, it is valid for laminar as well as for turbulent flows.

The coupling mechanisms between chemical reaction, flow, and transport processes (see Fig. 2) can be understood best if one adopts a Lagrangian viewpoint, where the coordinate system is fixed to volume elements moving with the flow [32]. First let us look at the coupling of convection (flow) with transport processes. Friction changes the velocity of the flow and provides a direct coupling mechanism. On the other hand convection (from a Lagrangian viewpoint) distorts the local coordinate systems, and, thus the local gradients in the flow. In this way it perturbs the transport processes (friction, heat conduction, diffusion, etc.). Furthermore, the transport processes directly change the thermokinetic state of the system, denoted by the vector ψ. Chemical reaction itself, changes ψ, too, and a strong coupling of chemical kinetics and molecular transport processes results. For high Mach number flows convection can change the enthalpy and the pressure, and couple with the chemical kinetics, but if we restrict to low Mach number flows, where the change of pressure and specific enthalpy due to velocity gradients is negligible, the chemical kinetics effects the flow through the density ρ, but there is no direct influence of the flow on the chemical kinetics. For turbulent flows this is reflected in the fact that turbulence influences the molecular transport processes via a building-up of gradients (through a distortion of the coordinate system), but that there is no direct chemistry-turbulence interaction — only an indirect one via molecular transport processes!

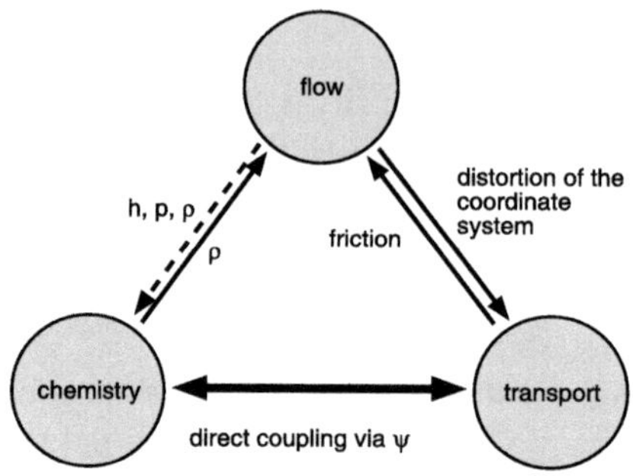

Fig. 2 Coupling of flow, chemistry, and transport processes.

3 Detailed Models

As described above, the enormous computational effort prohibits the use of detailed simulations (detailed kinetics, detailed transport models, resolution of the smallest time scales and the smallest spatial structures) for technical combustion devises [35]. Therefore modeling assumptions for the chemical kinetics, the turbulence-chemistry interaction and the multi-phase processes have to be made. However, many of the models used in current CFD-codes over-simplify the complex processes. On the other hand several tools exist which allow a detailed simulation of combustion processes with a minimum of modeling assumptions. Examples are:

- detailed simulations of the chemical kinetics based on comprehensive mechanisms [34, 35]
- direct numerical simulations of turbulent combustion processes [28, 3, 23]
- detailed simulations of multi-phase processes like ignition of droplets [9, 1, 30, 31]

Such detailed simulations reveal interesting insight in the underlying physical and chemical processes, but are computationally prohibitive for technical systems. In the following it will be shown how a hierarchical modeling strategy can overcome this problem. Based on detailed simulations it is possible to derive simplified models, which can then be used as submodels for realistic technical processes. In this way information from the smallest time- and length scales enters the global models.

3.1 Chemical kinetics

During the last years enormous progress has been made in the field of detailed kinetic modeling. Reliable kinetic mechanisms exist, which allow a reliable simulation of flame propagation and even of low-temperature ignition processes (see, e.g., [35] for references). Such mechanisms include the important reaction paths like the attack of the hydrocarbons by radicals and the oxidation of the hydrocarbons. The problem is however, that these mechanisms comprise in some cases several thousand chemical species and several thousand chemical reactions, yielding very large systems of stiff differential equations that have to be solved.

3.2 Chemistry-Turbulence Coupling

Turbulent combustion processes are governed by a strong interaction of chemistry, molecular transport and flow (see discussion above, section 2, Fig. 2). Furthermore a broad spectrum of length-, time-, and velocity scales renders the numerical simulation very difficult [16]. Direct numerical simulations [27, 3, 23, 33] are in principle possible (see Fig. 3), but computationally prohibitive for general technical combustion processes. Nevertheless they allow important insight into the interaction between molecular transport, chemical reaction, and flow processes. As an example, Fig. 3 shows the structure of a turbulent premixed H_2O_2, and H_2O flame [33]. An unperturbed laminar flame in a two-dimensional domain is assumed as a starting point, where on the left side ($x = 0$) in Fig. 3 we have the cold unburned premixed gas mixture and at the right boundary hot equilibrium conditions. In the middle of the x-axis the flame front is located, which can be easily seen from the mass fraction profile of H_2O_2. This species is produced and consumed rapidly in the flame front which makes it a good indicator for the flame front. Then an initial turbulence field is imposed. The turbulence intensity is $u' = 26.36\,\mathrm{m/s}$, the integral length $l_0 = 1.12\,\mathrm{mm}$, and the turbulent Reynolds number $Re_\mathrm{t} = 300$.

The flame front is perturbed by the turbulence field, which can be observed by the wrinkles in Fig. 3 for a calculation time of $t = 8\,\mu\mathrm{s}$. Later in time the perturbation of the flame front increases. At $t = 22\,\mu\mathrm{s}$ one might observe a beginning island formation, which shows the increasing influence of the flow field on the structure of the flame.

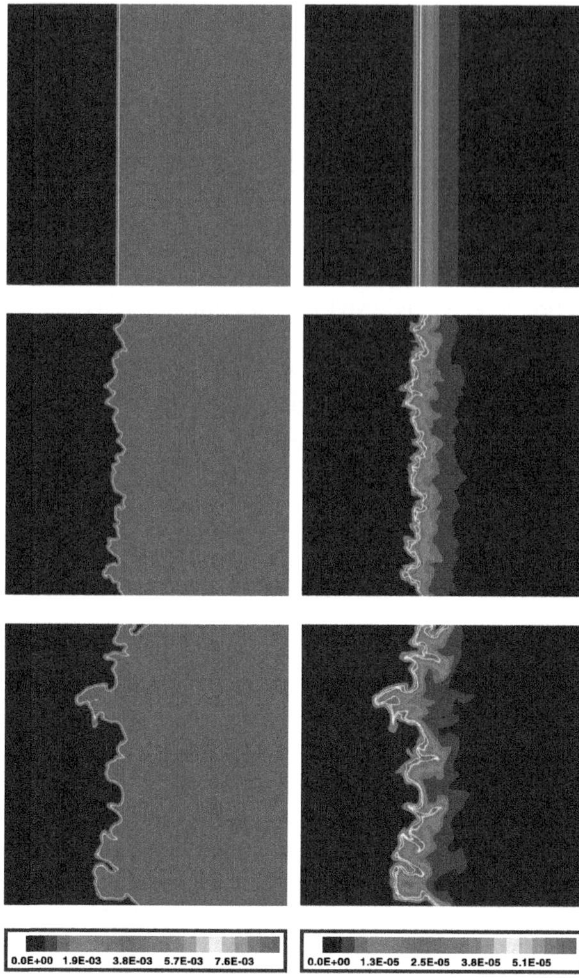

Fig. 3 Contour lines of a flame of a H_2/ air mixture at $t_{ac} = 0\,\mu$s (upper row), at $t_{ac} = 8\,\mu$s (middle row) and $t_{ac} = 22\,\mu$s (lower row); left — OH, right — H_2O_2.

3.3 Modeling of Multi-Phase Processes

Detailed simulations of droplet ignition can been performed by solving the coupled system of governing equations for liquid and gas phase. In this way, e.g. the influence of a locally inhomogeneous flow on the ignition process of droplets (important for spray ignition in Diesel engines) has been investigated [30, 31]. As an example combustion of a methanol droplet is shown in Fig. 4. With increasing flow velocity, the shape of the flame deviates more and more from a spherical shape. The extension of the flame in x-direction increases with increasing gas velocity. At the same time the extension in y-direction becomes smaller. Furthermore, it is remarkable that the maximal OH-concentration increases with increasing flow velocity. The maximum OH-concentration is located at the stagnation point upstream of the droplet. On the other hand the maximum OH-concentration downstream of the droplet decreases with increasing flow velocity. As can be seen in Fig. 4 the droplet is surrounded by a non-premixed flame. As another example Fig. 5 shows, e.g. the dependence of

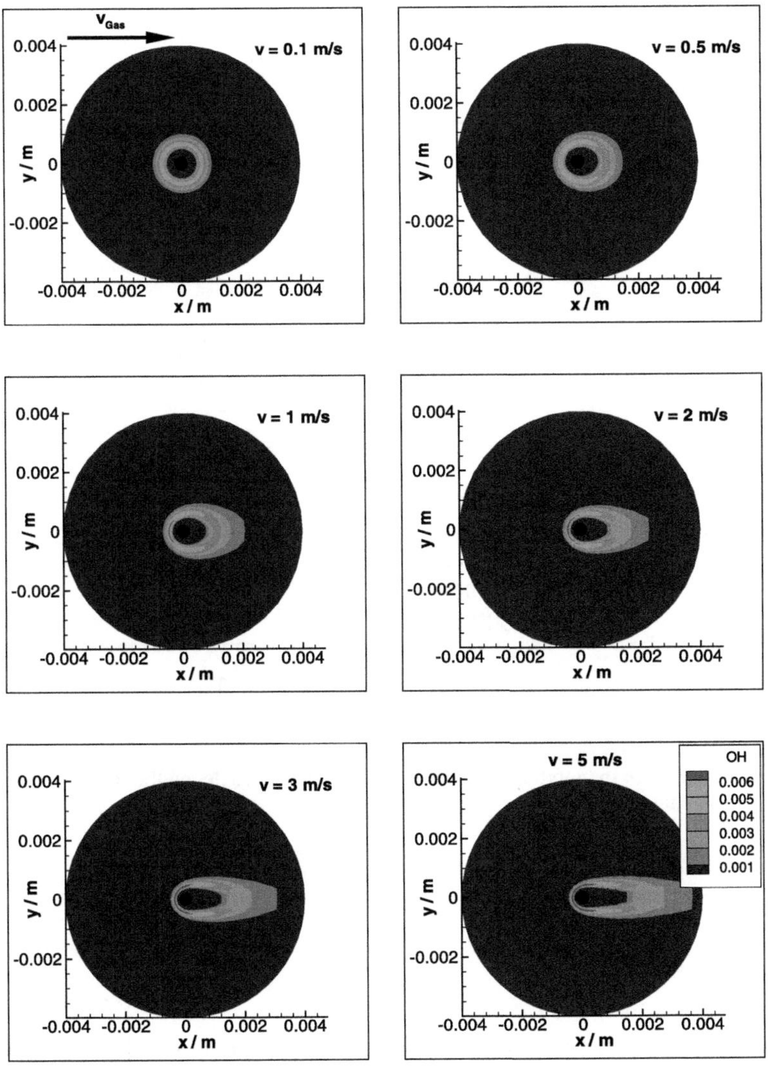

Fig. 4 Change of the flame structure with increasing gas velocity (methanol, T_g=1400 K, p=7 bar, $r_D(t=0)$=200 μm, t=9 ms, flow from the left to the right)

ignition delay time of a methanol droplet with 200 μm radius at 7 bar and different ambient temperatures on the velocity of the droplet relative to the gas phase, represented in terms of the Reynolds number, where the Reynolds number is defined as: $\mathrm{Re} = \rho_{air} v_{\mathrm{air}} \times d_D / \mu_{\mathrm{air}}$ with d_D as droplet diameter, and ρ_{air}, v_{air}, and μ_{air} as density, velocity, and viscosity of the ambient air, respectively. It can be seen that the ignition delay time decreases with increasing droplet velocity, which can be attributed to the increased local gradients and thus the improved mixing of fuel and air. For high droplet velocities, however, no ignition occurs. A significant decrease of the ignition delay time with increasing flow velocity can be observed. As one can see a variation of the flow velocity by an order of magnitude has a comparable effect on the ignition delay time like a variation of the gas temperature by ± 100 K. An understanding of this behavior is important e.g. for a reliable description of ignition processes in Diesel engines.

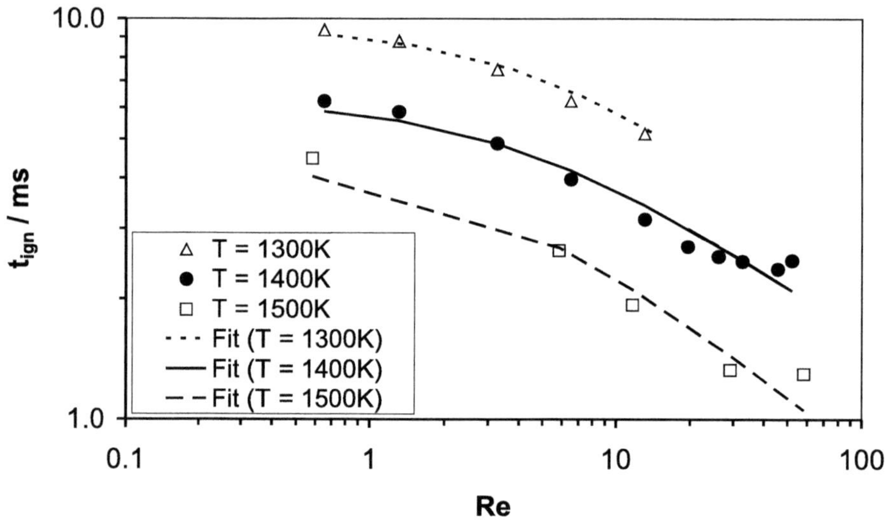

Fig. 5 Dependence of the ignition delay time for a methanol droplet on the Reynolds number (p = 7 bar, $r_D(0) = 200\,\mu$m)

4 Model Reduction

As described above, the use of detailed models is computationally prohibitive for a realistic technical combustion system. Therefore reduced models have to be devised which extract the important information from the detailed models such it can be used as sub-models for the overall global model.

4.1 Chemical Kinetics

Even though turbulence exhibits a strong interaction with the chemical kinetics, the dynamics of the processes shows a very simple structure, which can be explained by Fig. 6. It shows results of a Direct Numerical Simulation (DNS) for a non-premixed hydrogen-air flame for an initial turbulent Reynolds number of 210 and a turbulent characteristic time τ_t of 0.536 ms [23]. In the upper left figure the initial temperature field is shown, whereas in the upper right figure the turbulent flame is shown a time $t = 2.2\tau_t$. It can be seen that the turbulent flow field perturbs the structure of the flame considerably. In the lower parts of Fig. 6 the results of the DNS have been represented as scatter plots in the state space (here we show only a 3-dimensional plot of the results with the mass fractions of N_2, H_2O, and H as coordinates). In the left figure it can be seen that there is a direct correlation between the different mass fractions, because the DNS has been initialized with the profiles of a one-dimensional flame, where all the mass fractions are unique functions of the mixture fraction (represented by the mass fraction of N_2). After the turbulent flow field has had enough time to perturb the flame, some of the correlations cease to exist. But the figure shows that still not the whole composition space is accessed and that the accessed states seem to be in the neighborhood of intrinsic low-dimensional manifolds [21].

It is possible to determine these low-dimensional manifolds by a correlation analysis of DNS data (see [23] for details). This involves, however, an enormous computational effort. Other strategies to identify the ILDMs are based on time scale analyses of the chemical kinetics [21, 18]. As an example, Fig. 7 shows results for a stoichiometric methane-air system. Plotted are the correlations between H_2O, CO_2, the temperature T and CO (contours) as well as between H_2O, CO_2, OH and the temperature (contours). The domain where these correlations exist is basically the domain where chemistry is important. Outside this domain other correlations are introduced by physical processes such as diffusion [19]. In these figures it can be seen that if we know the concentrations of H_2O and CO_2, we explicitly know the temperature and the OH-concentration. In addition such figures are obtained for all other thermokinetic state variables (species concentrations, density, etc., not shown here), and therefore the thermokinetic state is known in this example in terms of the two reaction progress variables (H_2O and CO_2).

Based on these results the large system of evolution equations for the thermokinetic state can be replaced by a much smaller system for the reaction progress variables θ. It has to be noted that the reduction of the system is performed by a projection of the conservation equations onto the smaller space of reaction progress variables according to a procedure described in detail in [20, 22, 17, 5, 2]. One problem of the ILDM concept is the fact that it is based on an analysis of the chemical source term only although it is known that there is a strong coupling of the kinetics with the molecular transport processes. A strategy which overcomes this problem is based on the construction of reaction-diffusion manifolds (REDIM) in composition space [6, 7].

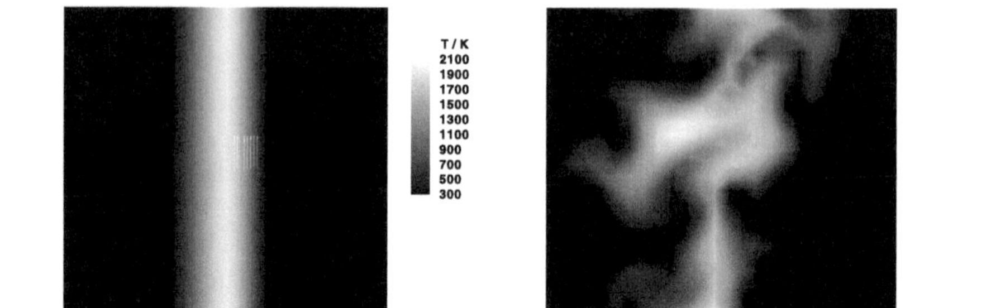

Fig. 6 Results of DNS of a hydrogen-oxygen-nitrogen non-premixed flame [23]. Upper figures: data plotted in the physical space (x-y-plots); Lower figures: scatter plots in the composition space. The left figures correspond to the initial conditions, the right figures to the results after $t = 2.2\tau_t$.

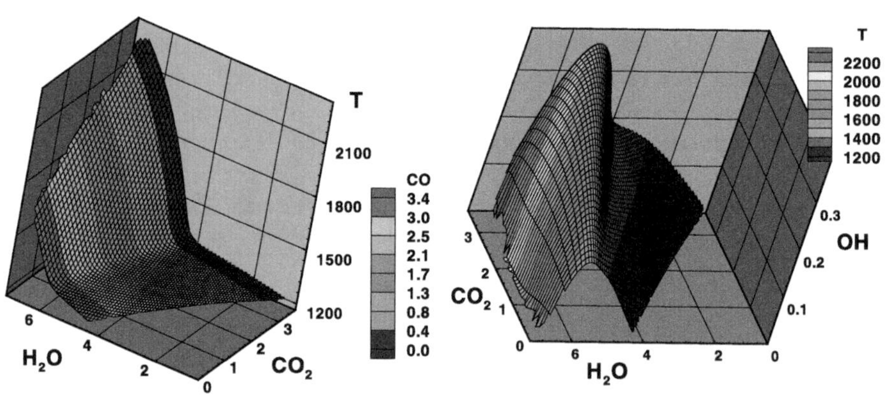

Fig. 7 Plot of two-dimensional ILDMs of the methane-air system in the state space. Concentrations are in units of mass fractions divided by the molar weight, temperature is in units of K.

This concept is based on a relaxation process, where an initial guess for a low-dimensional manifold evolves in such a manner that an invariant slow reaction/diffusion manifold is obtained. One major advantage of the REDIM technique over the ILDM concept is the fact that a REDIM exists in the whole accessed domain, even at low temperatures (close to the unburned mixture), where chemistry is slow and ILDM does not yield an existing manifold. Close to equilibrium the REDIM is typically close to the ILDM. It has been shown in [6] that, if chemistry governs the overall process, i.e. gradients tend to zero, the REDIM concept yields slow manifolds or, equivalently, iteratively refined ILDMs as a limiting case [25]. The information that enters the model is only the detailed mechanism and an estimate for the local gradients in the flames. The dependence on the gradients has been studied in [7] for the syngas(N_2, H_2, CO)/air system. It was shown that the manifold depends weekly on different approximations of the system gradients (see Fig. 8). Figure 8d shows that for three approximations which vary by one order of magnitude one from another (red, blue and green meshes) there are only minor differences between the approximations. In all other projections (a,b,c) the manifolds are generally the same with perturbations in the region close to the unburned point, where some deviations on the boundary appear [7].

4.2 Chemistry-Turbulence Interaction

A direct numerical simulation of the conservation equations is in principle possible, but computationally prohibitive for technical systems. Furthermore, the large parametric sensitivity would require averaging over a large number of such simulations in order to obtain the relevant properties like average temperature, average pollutant concentration. Due to the non-linearity of the chemical source term completely wrong results are obtained if the average chemical source term is calculated based on the averages of temperature and species concentrations [35, 16]

$$\bar{\omega}_i \neq \omega_i(\bar{\psi}) \qquad \psi = (h, p, w_1, \ldots, w_{n_s})^T \; . \tag{7}$$

Therefore statistical models are used which based on the joint probability density function (PDF) of velocity and scalars

$$f(\psi, V; x, t)\,\mathrm{d}\psi\,\mathrm{d}V = \mathrm{Prob}\{\psi \le \phi(x,t) < \psi + \mathrm{d}\psi, V \le U(x,t) < V + \mathrm{d}V\} \; . \tag{8}$$

The PDF represents the probability that the scalars $\phi(x,t)$ are in the interval $\psi, \psi + \mathrm{d}\psi$ and the velocity vector $U(x,t)$ in $\mathbf{V}, \mathbf{V} + \mathrm{d}\mathbf{V}$. If this PDF is known, the chemical source terms can be calculated via

$$\bar{\omega}_i = \int_{-\infty}^{+\infty}\int_{-\infty}^{+\infty} \omega_i(\psi)\, f(\psi, V; x, t)\,\mathrm{d}\psi\,\mathrm{d}V \; . \tag{9}$$

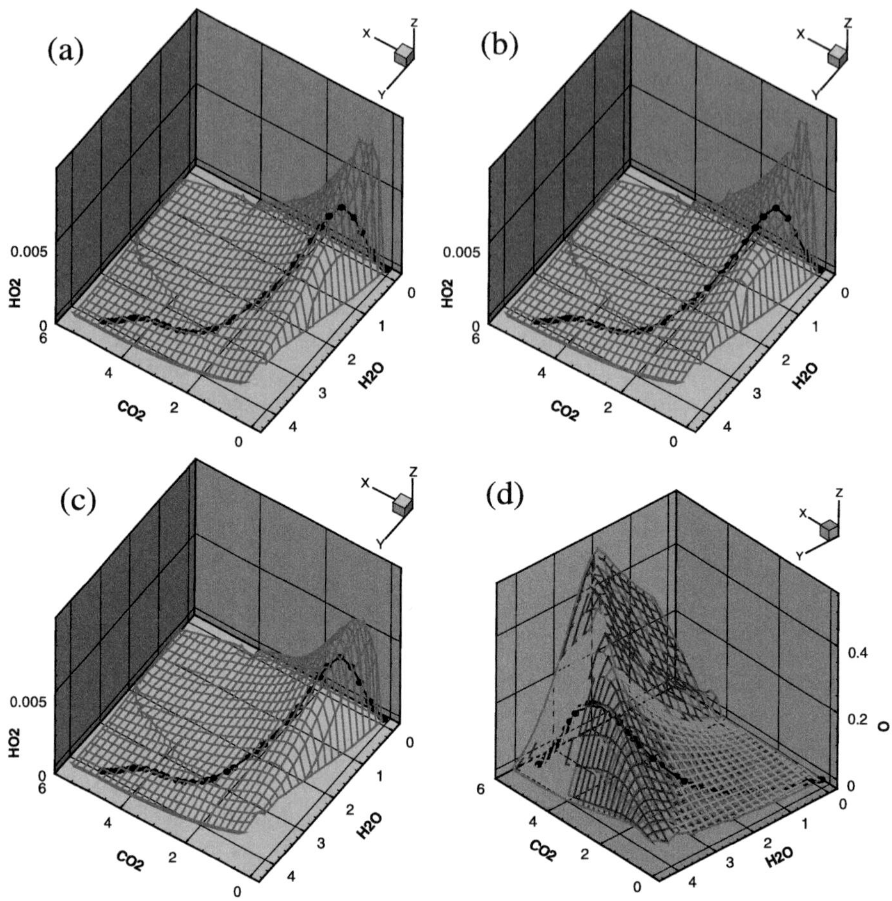

Fig. 8 Dependence on the gradient's approximation for 2D-reduction model, 2D-manifolds. **a** the gradients approximation has the order $\approx 10^4$ **b** 10^3, **c** 10^5 for minor species HO_2, **d** three cases (a,b,c) together (*red, blue, green*) for major species O and the detailed stationary solution (*solid black line*) .

Note that because the chemical source terms depend on the scalars only and not their gradients, the knowledge of the PDF is sufficient to calculate the average chemical source terms.

One possibility to calculate the time- and space-dependent PDF is to assume a generic shape and solve balance equations for the moments (average, variance, covariances) [16, 26, 35]. The problem is the assumption of the shape and the assumption of statistical independence of the different variables in order to represent the PDF as a product of marginal PDFs. An accurate method is to solve a transport equation for the PDF [12, 10]. In this method the chemical kinetics are treated in closed form, but molecular transport still has to be modeled, because the PDF does not contain information of two-point processes [15]. However, information from DNS on species and gradient correlations [23] can be used for closing the model. A major problem is the high dimensionality of the transport equation, because each chemical species enters as an independent variable. Therefore typically Monte-Carlo-methods are used for the simulation [12].

4.3 Multi-Phase Processes

Based on the formulation of a transport equation for the joint PDF of gas-phase and particle properties by Zhu et al. [36], a particle method has been developed for the solution of this high dimensional partial differential equation. In order to simplify the numerical treatment, the PDF of the gas phase and the particle phase have been formulated and discretized separately. As a first step only three properties of the liquid phase have been accounted for, namely the radii of the drops, their temperature and their velocity. This leads to a discretized formulation of the spray PDF according to

$$f(\mathbf{x},\mathbf{V},R,T,t) \approx \frac{1}{N_p}\sum_{n=1}^{N_p} \delta\,(\mathbf{x}-\mathbf{x}^{(n)}(t))\delta\,(\mathbf{V}-\mathbf{V}^{(n)}(t))$$
$$\delta\,(R-R^{(n)}(t))\delta\,(T-T^{(n)}(t))$$

where $\mathbf{x}$ denotes the location of droplet, $\mathbf{V}$ its velocity, R its radius, and T its temperature. N_p denotes the overall number of stochastic particles for the disperse phase and t the time.

The droplet velocity is modeled accounting for friction with the gas phase according to the average velocity of the gas phase and a random fluctuation velocity. Future improvements will focus on improving this model by taking into account the PDF of the gas phase velocity. The change of the particle radius is described by a d^2-law with a correction for the drift velocity. For the temperature of the droplets a simple model has been employed which accounts for the heating of the droplet. The thermokinetic state is calculated via a flamelet-like library of droplet ignition and burning scenarios, which has been generated using detailed simulations [30, 31]. The pressure field is calculated via the solution of a Poisson equation. Because of the statistical noise stemming from the particle representation, smoothing techniques had to be employed. The results will be presented in the next section.

5 Coupling of the Sub-Models

In the previous sections, it was described how the information from detailed simulations of chemical kinetics, droplet combustion, and chemistry-turbulence interaction can be used to devise reduced models. In order to simulate technical processes these

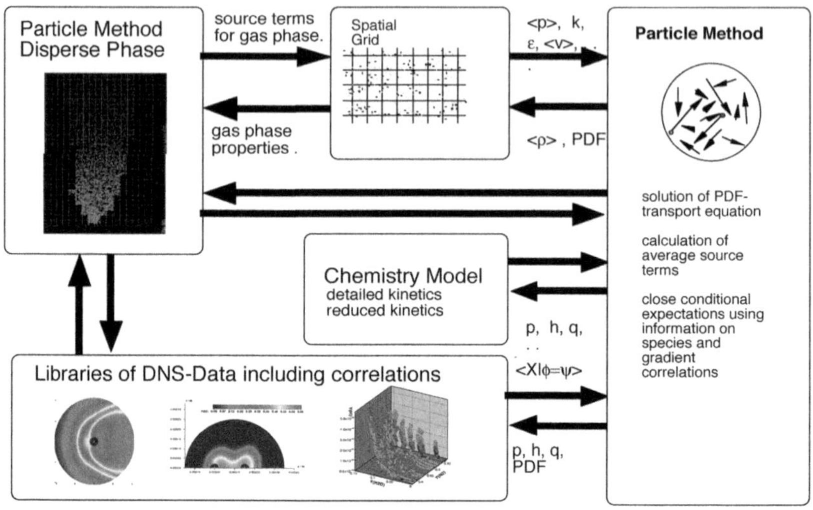

Fig. 9 Coupling of the sub-models

sub-models have to be coupled. One coupling strategy is shown in Fig. 9. It is based on different components.

Central is a statistical model which is based on the solution of a transport equation for the PDF of the gas phase. Two different approaches can be used to obtain the mean pressure field, namely either via the solution of a Poisson equation or via a coupling with a standard CFD-code. The PDF model for the gas phase yields information on the density fields which in turn is used to calculate the velocity field. Information on the chemistry and turbulent mixing is obtained from the sub-models for the chemistry-turbulence interaction and the kinetics.

The overall model allows a reliable simulation of turbulent combustion processes such as non-premixed turbulent natural gas flames, spray flames, and engine combustion. Sample results for gases flames can be found e.g. in [24]. First validations of the model for spray flames have been carried out by comparison of computational results of a methanol jet flame, which has been investigated experimentally by Chen et al. [8]. These experiments have been chosen, because they provide a good stationary test case. For this model flame, a flamelet representation of the processes in the gas phase has been chosen in order to simplify the treatment. Figure 10 shows as an example calculated mean scalar fields of a methanol spray flame investigated experimentally by Masri et al.. A quantitative comparison is obtained e.g. in Fig. 11, where radial profiles of the mean temperature are compared with experimental results.

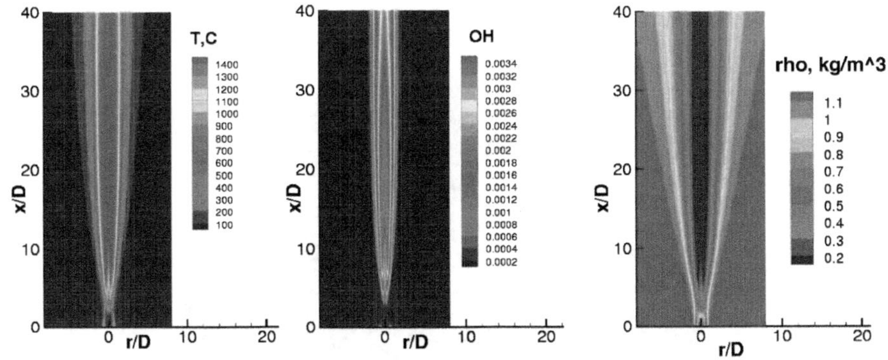

Fig. 10 Calculated temperature, OH and density fields of the Methanol spray flame [29].

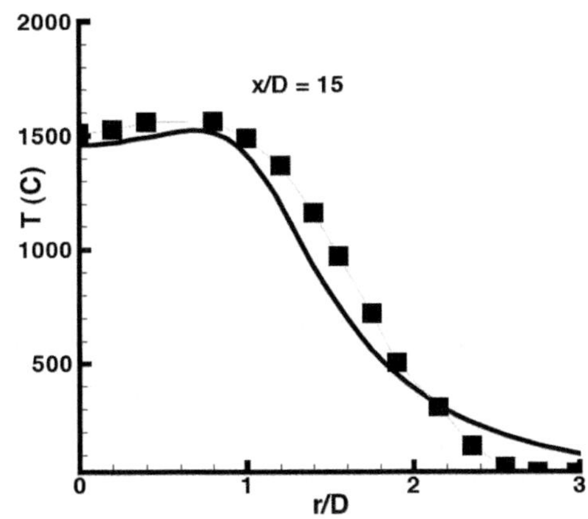

Fig. 11 Experimental [8] (*symbol*) and calculated radial profiles of temperature at one axial location ($x/D = 15$) [29].

6 Summary

The examples presented above show how detailed simulations in combination with hierarchical modeling strategies can be used efficiently to model technical combustion systems. Based on detailed simulations reduced models for the chemical kinetics, turbulence/chemistry interaction, and multiphase processes can be devised which are then coupled to a global model for the overall combustion process (see Fig. 12).

Acknowledgements Several of the simulations were performed on the national super computer NEC SX-8 at the High Performance Computing Center Stuttgart (HLRS) under the grant number flame2D/12792. The authors acknowledge the continuing support by the HLRS.

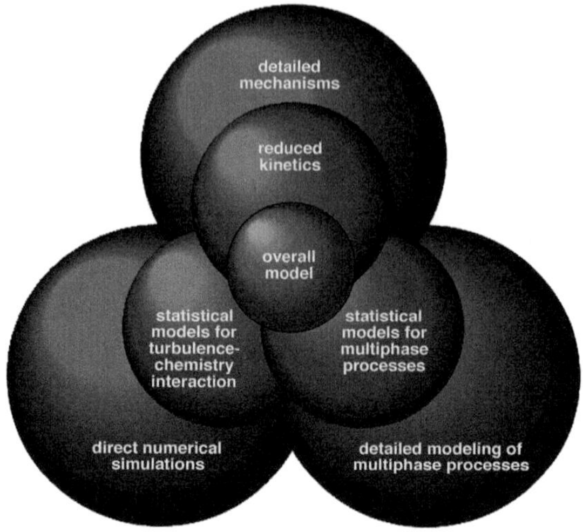

Fig. 12 Coupling of the sub-models 2

References

1. Aouina, Y., Gutheil, E., Maas, U., Riedel, U., Warnatz J.: Mathematical Modeling of Droplet Heating, Vaporization, and Ignition Including Detailed Chemistry. Comb. Sci. Tech. **173**, 91–114 (2001)
2. Bauer, J., Bykov, V., Maas, U.: Implementation of ILDMs based on a representation in generalized coordinates. In: P. Wesseling, E. Onate, J. Periaux (eds.) Proc. European Conference on Computational Fluid Dynamics, Egmond aan Zee (2006)
3. Baum, M.: Direct Numerical Simulation — A tool to study turbulent reacting flows. *Annual Reviews of Computational Physics*, D. Staufer (ed.), vol. 5, World Scientific Publishing Company (1997)
4. Bird, R., Stewart, W., Lightfoot, E.: Transport Phenomena, Wiley Interscience, New York (1960)
5. Blasenbrey, T., Schmidt, D., Maas, U.: Automatically Simplified Chemical Kinetics and Molecular Transport and its Application in Premixed and Non-Premixed Laminar Flame Calculations. Proc. Comb. Inst. **28**, 505–511 (1995)
6. Bykov, V., Maas, U.: Extension of the ILDM method to the domain of slow chemistry. Proc. Comb. Inst. **31**(1), 465–472 (2007)
7. Bykov, V., Maas, U.: The extension of the ILDM concept to reaction-diffusion manifold. Comb. Theory and Modell. **11**(6), 839–862 (2008)
8. Chen, Y., Stårner, S., Masri, A.R.: Further Study of Spray Combustion in a Simple Turbulent Jet Flow. In: Proc. 15th Australasian Fluid Mechanics Conference, Sydney (2004)
9. Cho, S., Yetter, R., Dryer, F.: A computer model for one-dimensional mass and energy transport in and around chemically reacting particles, including complex gas-phase chemistry, multicomponent molecular diffusion, surface evaporation, and heterogeneous reaction. J. Comput. Phys. **102**, 160–179 (1992)
10. Dopazo, C., O'Brien, E.: An approach to the autoignition of a turbulent mixture. Acta Astronaut **1**, 1239–1266 (1974)
11. Ern, A., Giovangigli, V.: Multicomponent Transport Algorithms, *Lecture Notes in Physics*, Springer, Berlin Heidelberg New York (1994)
12. Haworth, D., Pope, S.B.: A generalized Langevin model for turbulent flows. Physics of Fluids **29**, 387–405 (1986)

13. Hirschfelder, J., Curtiss, C.: Theory of Propagation of Flames. Part I: General Equations. In: Proc. 3rd Symposium (International) on Combustion, Williams and Wilkins, Baltimore (1949)
14. Hirschfelder, J., Curtiss, C.: Molecular Theory of Gases and Liquids. John Wiley & Sons Inc., New York (1964)
15. Juneja, A., Pope, S.B.: A DNS study of turbulent mixing of two passive scalars. Physics of Fluids **8**, 2161–2184 (1996)
16. Libby, P., Williams, F.: Turbulent Reacting Flows. Academic Press, New York (1994)
17. Maas, U.: Coupling of Chemical Reaction with Flow and Molecular Transport. Applications of Mathematics **3**, 249–266 (1995)
18. Maas, U.: Efficient Numerical Calculation of Intrinsic Low-Dimensional Manifolds in Composition Space. Computing and Visualization in Science **1**(2), 69–81 (1997)
19. Maas, U.: Mathematical Modeling of the Coupling of Chemical Kinetics With Flow and Molecular Transport. *Scientific Computing in Chemical Engineering II*, Springer (1999)
20. Maas, U., Pope, S.B.: Implementation of Simplified Chemical Kinetics Based on Intrinsic Low-Dimensional Manifolds. Proc. Comb. Inst. **24**, 103–112 (1992)
21. Maas, U., Pope, S.B.: Simplifying Chemical Kinetics: Intrinsic Low-Dimensional Manifolds in Composition Space. Comb. Flame **88**, 239–264 (1992)
22. Maas, U., Pope, S. B.: Laminar Flame Calculations using Simplified Chemical Kinetics Based on Intrinsic Low-Dimensional Manifolds. Proc. Comb. Inst. **25**, 1349–1356 (1994)
23. Maas, U, Thévenin, D.: Correlation Analysis of Direct Numerical Simulation Data of Turbulent Non-Premixed Flames. Proc. Comb. Inst. **27**, 1183–1189 (1998)
24. Merci, B., Naud, B.D., Roekaerts, D., Maas, U.: Joint Scalar versus Joint Velocity-Scalar PDF Simulations of Bluff-Body Stabilised Flames with REDIM. Flow, Turbulence and Combustion, in print (2008)
25. Nafe, J., Maas, U.: A General Algorithm for Improving ILDMs. Comb. Theory and Modell. **6**(4), 697–709 (2002)
26. Peters, N.: Turbulent Combustion. MIT Press, Cambridge (2000)
27. Poinsot, T., Candel, S., Trouve, A.: Applications of direct numerical simulation to premixed turbulent combustion. Prog. Energy Combust. Sci. **21**, 531–576 (1996)
28. Poinsot, T., Haworth, D., Bruneaux, G.: A law of the wall model for turbulent premixed combustion. Comb. Flame **95**, 118–133 (1993)
29. Rybakov, A., Maas, U.: In preparation (2008)
30. Stauch, R., Maas, U.: The auto-ignition of single *n*-heptane/*iso*-octane droplets. Int. J. Heat and Mass Transfer **50**, 3047–3053 (2007)
31. Stauch, R., Maas, U.: The ignition of methanol droplets in a laminar convective environment. Comb. Flame **153**, 45–57 (2008)
32. Taylor, G.: Diffusion by continuous movement. Proc. Lond. Math. Soc. **20**, 196–212 (1921)
33. Tsai, W., Schmidt, D., Maas, U.: Direct Numerical Simulations of Spark Ignition of H_2/Air-Mixture in a Turbulent Flow. In: E. Krause, W. Jäger (eds.) Proc. High Performance Computing in Science and Engineering 2000, Springer-Verlag Berlin Heidelberg, 433–441 (2001)
34. Warnatz, J.: Resolution of Gas Phase and Surface Chemistry into Elementary Reactions. Proc. Comb. Inst. **24**, 553–579 (1993)
35. Warnatz, J., Maas, U., Dibble, R.: Combustion. 4 edn., Springer-Verlag, Berlin Heidelberg (2004)
36. Zhu, M., Bray, K.N.C., Rumberg, O., Rogg, B.: PDF Transport Equations for Two-Phase Reactive Flows and Sprays. Comb. Flame **122**, 327–338 (2000)

Understanding Molecular Recognition and Self-Assembly from Large-Scale Numerical Simulations

Stephan Blankenburg and Wolf Gero Schmidt

Abstract Nowadays, complex chemical problems such as the origin and mechanism of molecular recognition and self-assembly can be addressed computationally, using high performance resources. This is illustrated in the following, using the adsorption of small amino acids and DNA base molecules on metals as an example. *First-principles* calculations are used to rationalize the long-range chiral recognition between adenine and phenylglycine adsorbed on Cu(110) [Chen and Richardson, Nature Materials 2, 324 (2003)]. The enantiomeric interaction is traced to substrate-mediated Coulomb repulsion and template effects. The mechanism revealed here (i) shows that the Easson and Stedman model for chiral recognition may include long-range electrostatic interactions and (ii) illustrates the catalytic potential of the substrate for molecular self-assembly.

1 Introduction

Molecular recognition and the self-assembly of molecular structures are ubiquitous in nature, but also increasingly used in chemical synthesis and nanotechnology. The mechanisms that underly these fascinating processes, however, are often poorly understood. Surface-adsorbed molecules are popular model systems to puzzle out the details of the molecular interactions [15, 8, 14, 17, 6, 10, 23]. Scanning tunneling microscopy (STM) studies on adenine and phenylglycine adsorbed on Cu(110) [5] revealed a particularly intriguing example of molecular recognition. The system is remarkable not only because the interplay of nucleic acid bases and amino acids is of fundamental importance for many biological processes. It is also the first direct observation of diastereoisomeric interactions due to chiral recognition between dissimilar molecules. Enantiomeric interactions are commonly explained within the

Stephan Blankenburg · Wolf Gero Schmidt
Lehrstuhl für Theoretische Physik, Universität Paderborn, 33095 Paderborn, Germany, e-mail: {blankenburg@phys.upb.de, w.g.schmidt@upb.de}

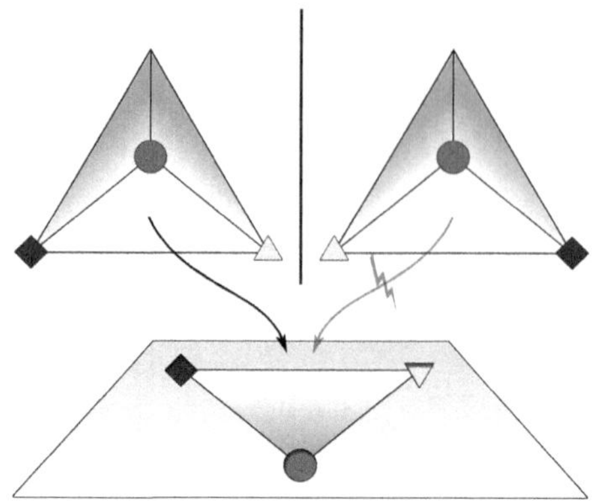

Fig. 1 (color online) Schematic illustration of the "three-point" contact model for chiral discrimination: The molecule on the left matches the three nonequivalent bonding sites, in contrast to its mirror-imaged enantiomorph on the right.

"three-point" contact model [7, 4], shown schematically in Fig. 1. In this model due to Easson and Stedman [7], stereochemical differences in reactivity are due to the differential bonding of enantiomers with three nonequivalent bonding sites. Discrimination occurs when one isomer can simultaneously interact with all three sites, while its enantiomorph cannot. However, in the case of adenine and phenylglycine co-adsorbed on Cu(110), the chiral discrimination acts at a distance of up to 20 Å [5], i.e., is seemingly beyond the "three-point" contact model of chiral recognition.

Let us briefly summarize the experimental findings. Chen and Richardson [5] observed that adenine deposited on Cu(110) at room temperature forms ordered one-dimensional molecular dimer chains that grow along the lateral [$\pm$1,2] directions (given with respect to the [$\bar{1}$10] and [001] Cu crystal orientations, see Fig. 2). Co-adsorbed phenylglycine shows a strong chiral preference in its interaction with these chains: S-phenylglycine attaches to [1,2] oriented chains, whereas R-phenylglycine decorates chains aligned along [-1,2]. The STM images show double rows of phenylglycine molecules that run parallel to the adenine dimer chains. The microscopic interpretation of the [1,2] oriented chain structure is shown in Fig. 2.

Here we rationalize this fascinating example of chiral recognition with the help of density functional theory (DFT) calculations. It is shown that the long-range enantiomeric interaction is mediated by the metal substrate. This (i) acts as a checkerboard that restricts the lateral degrees of freedom of the admolecules and (ii) enables charge accumulation at the admolecules leading to long-range Coulomb forces.

2 Computational Method

The calculations are performed using the Vienna Ab Initio Simulation Package (VASP) implementation [12] of DFT, using the PW91 functional [21] to model electron exchange and correlation. The electron-ion interaction is described by the projector-augmented wave (PAW) method [13], which allows for an accurate treatment of the first-row elements as well as the Cu 3*d* electrons with a relatively moderate energy cutoff of 340 eV. The surface Brillouin zone is sampled using a $2\times2\times1$ mesh. The adsystem is modeled by periodically repeated slabs, containing six atomic Cu layers plus the adsorbed molecules and a vacuum region equivalent in thickness to about 17 atomic Cu layers. This methodology was found to reproduce the measured geometries for phenylglycine [2] and adenine [22] adsorbed on Cu(110). We mention that the PW91 functional allows for a reasonable description of hydrogen bonds, at least in the case of solid water [9, 24]. In order to estimate the H-bond strength within Bader's topological paradigm [1], we use a functional of the calculated charge density [11]. Typical calculation times vary from one day up to two weeks (wall-clock time) using 32 CPUs on a NEC SX-8 system. For such a number of CPUs we reach an overall speedup of about 18 as demonstrated in Fig. 3. Compared to the CRAY opteron cluster strider the calculations are substantially faster (cf. Fig. 3) and more memory can be accessed.

3 Results and Discussion

The adenine-Cu(110) interaction is governed by mutual polarization and Coulomb attraction [23, 22]. The adsorption of phenylglycine on Cu(110), on the other hand, leads to covalent bonding [2]. Despite this difference, the adsorption characteris-

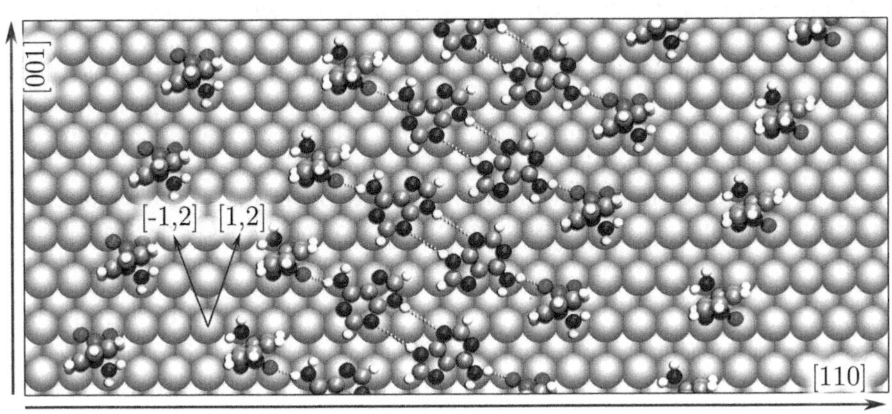

Fig. 2 Molecular model derived in [5] for phenylglycine co-adsorbed with adenine forming dimer rows along the [1,2] direction on Cu(110). Hydrogen bonds are indicated with white lines.

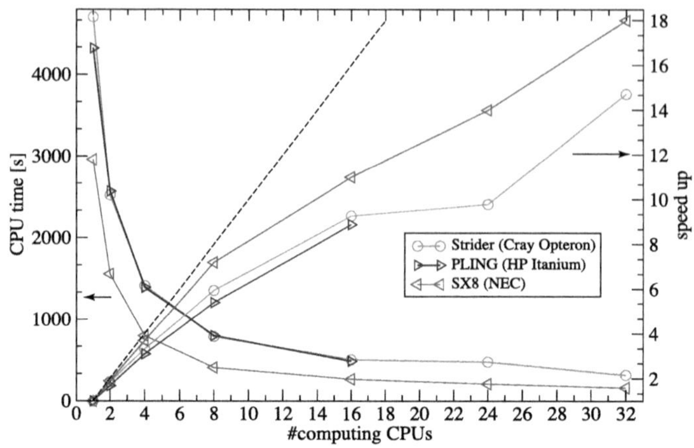

Fig. 3 CPU time and speedup versus the number of CPUs for a typical electronic structure calculation using VASP (see text). A NEC SX-8 (HLRS Stuttgart), a CRAY opteron (HLRS Stuttgart) and a HP Itanium II (University of Paderborn) cluster are compared.

tics of adenine and phenylglycine share one feature that is important in the present context: The energy barriers that hinder lateral movements of the molecules on the Cu(110) surface substrate are considerable, up to 0.5 and 1.0 eV for adenine and phenylglycine, respectively. In the present case, the lateral constraints are strengthened by hydrogen bonds between the carboxyl group of the first-row phenylglycine molecules and the adenine amino group, as well as within the adenine dimers, see Fig. 2. Hydrogen bonding in conjunction with steric constraints was found to be important for the enantiospecific interaction of glycine and phenylglycine on Cu(110) [19, 3]. Steric constraints, however, seem unlikely to be important here: The separation between the adenine chain and the nearest phenylglycine is about 1.15 nm along the [110] direction [5]. This distance roughly doubles in case of the second phenylglycine row.

The assumption that the [$\pm$1,2] direction imposed by the adenine chains acts as an enantiomeric selector is possibly the simplest hypothesis to explain the chiral recognition. In other words, one could suspect that the adenine dimer chain provides a template that forces the phenylglycine molecules to line up in a given direction. Due to, e.g., higher electrostatic moments or substrate mediated strain effects, the arrangement along [$\pm$1,2] may be more or less favorable for a given enantiomer. In order to probe this hypothesis, we perform calculations for S- and R-phenylglycine in a monoclinic supercell the basal plane of which has $\begin{pmatrix} 1 & 2 \\ 5 & 0 \end{pmatrix}$ periodicity. This forces the amino acids to assume the same translational symmetry as given by the adenine dimer chains oriented along [1,2]. The molecules are allowed to fully relax.

A number of different initial positions were probed and a rotational profile was calculated to optimize the phenyl-ring position. The calculations were performed with and without Cu substrate. A top view of the former case for S- and R-phenylglycine is shown in Figs. 4(a) and (b), respectively. The adsorption geometry agrees with earlier findings [2]. Irrespective of the presence of the substrate, the calculations find an energetic preference of one enantiomer, namely R-phenylglycine, for the given translational symmetry. The calculated energy difference $\Delta E_{s-r} = E_s - E_r$, however, is very small, 0.01 eV. More important, the preference of R over S-phenylglycine for the symmetry probed is in contrast to the experimental observation that S- rather than R-phenylglycine decorates [1,2] oriented adenine chains.

Obviously, symmetry constraints imposed on single rows of amino acids are not sufficient to explain the enantiospecific adsorption. Actually, the STM data show double rows of phenylglycine molecules parallel to the adenine dimer chains. The molecules farther away from adenine are found to be rotated by 180° with respect to the amino acid in the vicinity of the nucleic acid base, see Fig. 2. In order to see if the second molecular row changes the adsorption energetics, two phenylglycine molecules with the same chirality but opposite orientations were studied in a surface unit cell of $\begin{pmatrix} 1 & 2 \\ 6 & 0 \end{pmatrix}$ periodicity. The calculated energy difference ΔE_{s-r} doubles to 0.02 eV per molecule. This is still rather small and favors the R- rather than the S-enantiomer, in contrast to the experiment.

The calculations so far show that in fact adenine — or at least molecule-specific functional groups — are crucial for the enantiomeric adsorption of phenylglycine. Therefore, we now study molecular rows of adenine and phenylglycine adsorbed on Cu(110). The respective model systems for S- and R-phenylglycine with $\begin{pmatrix} 1 & 2 \\ 10 & 0 \end{pmatrix}$ periodicity are shown in Figs. 4(c) and (d). The dimension of the unit cell in [1,2] direction corresponds to the experiment, while the cell size in $[\bar{1}10]$ direction was gradually increased until the energy difference ΔE_{s-r} was converged. Again, a variety of starting configurations were probed and the rotational profile of the phenyl ring was sampled in order to verify that the ground state of the adsorption structure is reached. The calculations yield an energy difference $\Delta E_{s-r} = -0.10\,\text{eV}$ that is (i) significantly larger than resulting from the symmetry constraints discussed above and (ii) favors S-phenylglycine attachment, i.e., reproduces the experimental preference. We mention that the energy difference is of the same order of magnitude than the 0.2 eV found responsible for the formation of homo- rather than heterochiral cysteine dimers on gold [14].

The calculations for the model systems shown in Fig. 4(c) and (d) thus yield an energy difference that is suitable to explain the experiment. But what causes this energy difference? First we investigate the impact of the substrate by repeating the calculations for the frozen molecular adsorption structures without the substrate. Interestingly, removing the substrate reduces the energy difference between the two chiralities to $\Delta E_{s-r,nosubstr} = 0.01$ eV, i.e., the adenine-phenylglycine interaction is significantly enantiomeric only in the presence of the (achiral) substrate.

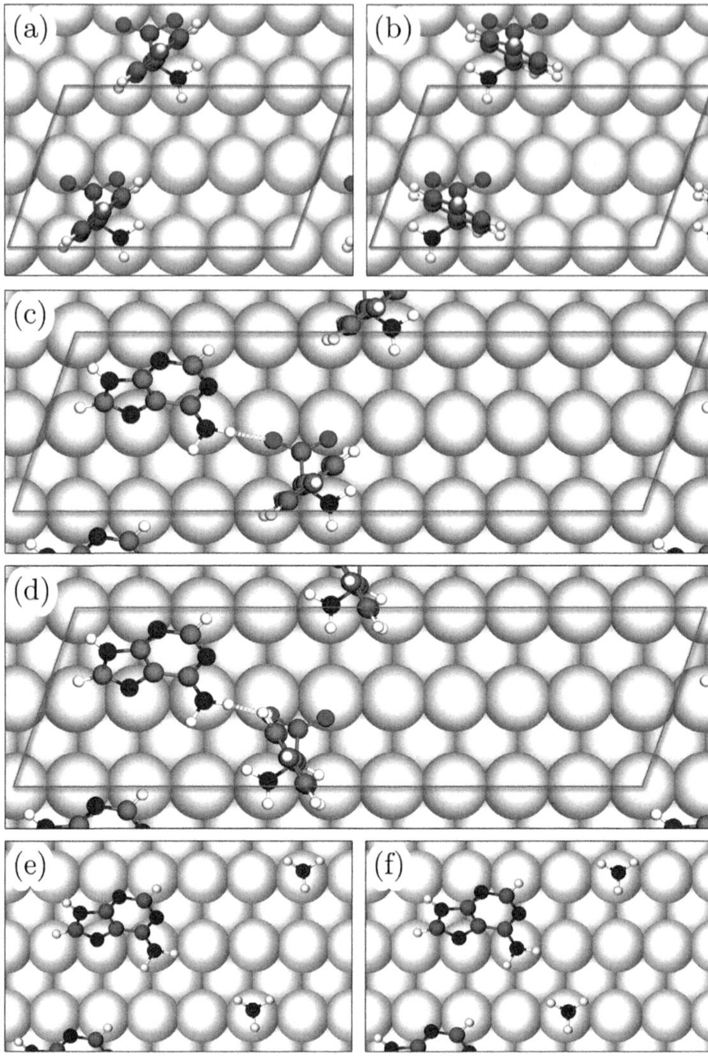

Fig. 4 (color online) S- **(a)** and R-phenylglycine **(b)** adsorbed on Cu(110) forced to form a row along the [1,2] direction. Molecular rows of S- **(c)** and R-phenylglycine **(d)** co-adsorbed with adenine on Cu(110). Red lines indicate the respective surface unit cells. Adenine and ammonia in the S- **(e)** and R-phenylglycine configuration **(f)** adsorbed on Cu(110)

The energy difference $\Delta E_{s-r} = -0.10\,\text{eV}$ can be broken down (following [3]) into its contributions from phenylglycine-adenine interaction $\Delta E_{s-r,inter} = -0.08\,\text{eV}$, molecule-substrate bonding $\Delta E_{s-r,bond} = -0.05\,\text{eV}$, and adsorption-induced strain energy $\Delta E_{s-r,strain} = 0.03\,\text{eV}$. Obviously, the adsorption of S- rather than R-phenylglycine parallel to adenine dimer chains along the [1,2] direction is mainly preferred due to more favorable adenine-phenylglycine interactions and somewhat more favorable molecule-substrate bonds, but involves slightly higher strain. Because the first contribution is the most important one, it will now be analyzed in detail.

One might suspect the hydrogen bond between phenylglycine and adenine of being responsible for the energy difference. However, within the approximation of the Bader approach [1, 11] and the numerical accuracy, we find no difference in the H-bond strength for the two enantiomers. Next, we partially decompose the amino acid in order to specify the functional group that is causing the energy difference for co-adsorbed R and S-phenylglycine. Replacing the phenyl group with hydrogen does not at all modify the interaction energy difference. Next, we study the energy differences for isolated carboxyl and amino groups that are frozen in the configurations they assume in the fully relaxed surface-adsorbed amino acid. While nearly no energy difference is calculated for the adenine-carboxyl group interaction, we find an appreciable difference for the hydrogen-saturated amino group, $\Delta E_{s-r,NH_3,inter} = -0.09\,\text{eV}$ (see model structures in Fig. 4(e) and (f)). It is of the same sign and magnitude as calculated for the complete adsystem.

This allows for deepening the analysis by simplifying the model system to the structures shown in Fig. 4(e) and (f). From the distance between ammonia and adenine we can exclude chemical interactions. To probe electrostatic interactions, we calculate the charge transfer between substrate and adsorbate. This is done by means of defining horizontal planes that cut through the center of the admolecule-substrate bonds. That procedure indicates a moderate and weak electron accumulation for ammonia and adenine, respectively: $Q_{s,NH_3} = -0.53e$, $Q_{r,NH_3} = -0.56e$, $Q_{s,adn} = -0.06e$ and $Q_{r,adn} = -0.07e$. From these values and the respective center of gravities for the charge we can estimate the electrostatic repulsion between adenine and ammonia using a point charge model [18]. The Madelung energy difference for the two ammonia positions that correspond to different phenylglycine enantiomers amounts to $\Delta E_{s-r,Coulomb} = -0.08\,\text{eV}$, i.e., agrees well with the difference of the respective total energies from the DFT calculation for the complete adsystems. We mention that the energy difference is mainly caused by the different NH_3-adenine distances, rather than by the slightly different charges. The effect is illustrated in Fig. 5. This mechanism does not work for the other functional groups of phenylglycine: The charge redistribution at the respective phenyl groups is too small and the orientation of the carboxyl groups is too similar for both the S and R adsorption configurations to cause a measurable enantiomeric interaction.

The fact that the substrate-adsorbate charge transfer causes the enantiospecific adsorption explains why the presence of the substrate is crucial for the chiral recognition of phenylglycine and adenine. However, the role played by the substrate is two-fold. Not only the substrate-adsorbate charge transfer is required for the enantiospecific interaction, also the locking of the adsorbate into specific adsorption sites due to the corrugation of the potential energy surface [22, 2] is essential. We mention that this limits the number of substrates which may be used for enantiomeric surface reactions of the kind discussed here. The weak corrugation of the potential energy surface and the small charge transfer found for adenine adsorption on graphite for example [20], will exclude similar observations for this very popular model substrate.

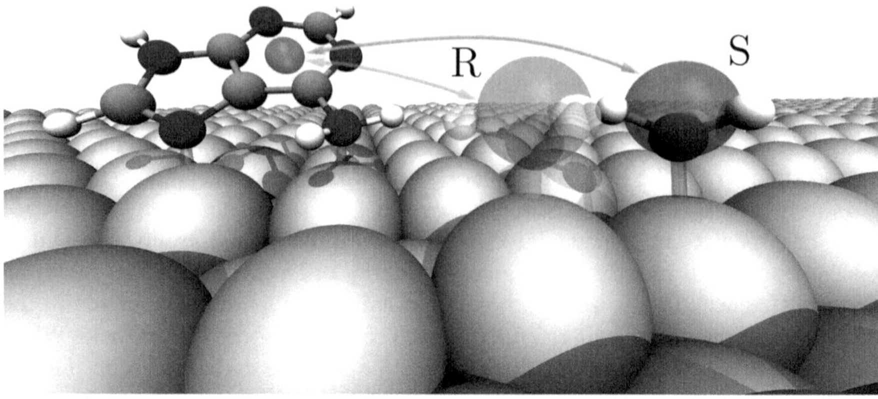

Fig. 5 (color online) Schematic illustration of the Coulomb interaction due to electron transfer from the substrate to the admolecules. For clearer presentation only one pair of charges is shown for ammonia in S- (full color) and R-phenylglycine configuration (shaded color).

4 Conclusions

In conclusion, we performed DFT calculations for adenine and S(R)-phenylglycine adsorbed on the Cu(110) surface. The calculated total energies are suitable to explain the experimental finding that S-phenylglycine decorates [1,2] oriented adenine chains, while [-1,2] oriented chains attract R-phenylglycine. By decomposing the amino acid in smaller building blocks we find the Coulomb repulsion between the phenylglycine amino-group and the DNA base to be responsible for the enantiospecific adsorption. The substrate-mediated charge transfer thus acts as chiral selector, while the direct intermolecular interactions such as hydrogen bonds do not. The calculations show (i) that electrostatic forces acting over large distances can constitute at least one of the interactions in the "three-point" contact model for enantioselectivity and (ii) that the substrate may in fact catalyze molecular recognition and self-assembly. For the complete adstructure observed experimentally — which due to its size still remains a challenge for large-scale calculations — further long-range interactions such as strain fields and charge-density waves [16] can be expected to additionally enrich the physics of the chiral recognition.

Acknowledgements We thank Neville Richardson for very helpful discussions. The calculations were done using grants of computer time from the Höchstleistungs-Rechenzentrum Stuttgart and the Paderborn Center for Parallel Computing (PC^2). The Deutsche Forschungsgemeinschaft is acknowledged for financial support.

References

1. Bader, R.F.W.: Atoms in molecules: A quantum theory. Oxford, USA (1990)
2. Blankenburg, S., Schmidt, W.G.: Adsorption of phenylglycine on copper: Density functional calculations. Phys. Rev. B **74**, 155419 (2006)
3. Blankenburg, S., Schmidt, W.G.: Steric effects and chirality in the adsorption of glycine and phenylglycine on Cu(110). Nanotechnology **18**, 424030 (2007)
4. Booth, T.D., Wahnon, D., Wainer, I.W.: Is chiral recognition a three-point process? Chirality **9**, 96 (1997)
5. Chen, Q., Richardson, N.V.: Enantiomeric interactions between nucleic acid bases and amino acids on solid surfaces. Nature Materials **2**, 324 (2003)
6. Di Felice, R., Selloni, A.: Adsorption modes of cysteine on Au(111): Thiolate, amino-thiolate, disulfide. J. Chem. Phys. **120**, 4906 (2004)
7. Easson, E.H., Stedman, E.: Studies on the relationship between chemical constitution and physiological action. Biochem. J. **27**, 1257 (1933)
8. Ernst, K.H., Kuster, Y., Fasel, R., Müller, M., Ellerbeck, U.: Two-dimensional separation of [7]helicene enantiomers on Cu(111). Chirality **13**, 675 (2001)
9. Hamann, D.R.: H_2O hydrogen bonding in density-functional theory. Phys. Rev. B **55**, R10,157 (1997)
10. Hauschild, A., Karki, K., Cowie, B.C.C., Rohlfing, M., Tautz, F.S., Sokolowski, M.: Molecular distortions and chemical bonding of a large π-conjugated molecule on a metal surface. Phys. Rev. Lett. **94**, 036106 (2005)
11. Jones, G., Jenkins, S.J., King, D.A.: Hydrogen bonds at metal surfaces: Universal scaling and quantification of substrate effects. Surf. Sci. **600**, L224 (2006)
12. Kresse, G., Furthmüller, J.: Efficiency of ab-initio total energy calculations for metals and semiconductors using a plane-wave basis set. Comput. Mater. Sci. **6**, 15 (1996)
13. Kresse, G., Joubert, D.: From ultrasoft pseudopotentials to the projector augmented-wave method. Phys. Rev. B **59**, 1758 (1999)
14. Kühnle, A., Linderoth, T.R., Hammer, B., Besenbacher, F.: Chiral recognition in dimerization of adsorbed cysteine observed by scanning tunneling microscopy. Nature **415**, 891 (2002)
15. Lorenzo, M.O., Baddeley, C.J., Muryn, C., Raval, R.: Extended surface chirality from supramolecular assemblies of adsorbed chiral molecules. Nature **404**, 376 (2000)
16. Lukas, S., Witte, G., Wöll, C.: Novel mechanism for molecular self-assembly on metal substrates: Unidirectional rows of pentacene on Cu(110) produced by a substrate-mediated repulsion. Phys. Rev. Lett. **88**, 028,301 (2002)
17. Nilsson, A., Pettersson, L.G.M.: Chemical bonding on surfaces probed by x-ray emission spectroscopy and density functional theory. Surf. Sci. Rep. **55**, 49 (2004)
18. Northrup, J.E., Froyen, S.: Structure of GaAs(001) surfaces: The role of electrostatic interactions. Phys. Rev. B **50**, 2015 (1994)
19. Nyberg, M., Odelius, M., Nilsson, A., Pettersson, L.G.M.: Hydrogen bonding between adsorbed deprotonated glycine molecules on Cu(110). J. Chem. Phys. **119**, 12577 (2003)
20. Ortmann, F., Schmidt, W.G., Bechstedt, F.: Attracted by long-range electron correlation: Adenine on graphite. Phys. Rev. Lett. **95**, 186101 (2005)
21. Perdew, J.P., Chevary, J.A., Vosko, S.H., Jackson, K.A., Pederson, M.R., Fiolhais, D.J.S.C.: Atoms, molecules, solids, and surfaces: Applications of the generalized gradient approximation for exchange and correlation. Phys. Rev. B **46**, 6671 (1992)
22. Preuss, M., Schmidt, W.G., Bechstedt, F.: Coulombic amino group-metal bonding: Adsorption of adenine on cu(110). Phys. Rev. Lett. **94**, 236102 (2005)
23. Schmidt, W.G., Seino, K., Preuss, M., Hermann, A., Ortmann, F., Bechstedt, F.: Organic molecule adsorption on solid surfaces: chemical bonding, mutual polarisation and dispersion interaction. Appl. Phys. A **85**, 387 (2006)
24. Thierfelder, C., Hermann, A., Schwerdtfeger, P., Schmidt, W.G.: Strongly bonded water monomers on the ice ih basal plane: Density-functional calculations. Phys. Rev. B **74**, 045,422 (2006)

Large Scale Particle-in-cell Plasma Simulation

Seiji Ishiguro

Abstract A three dimensional particle-in-cell (PIC) simulation code is developed for investigating plasma kinetic phenomena in fusion plasma. The code is parallelized using High Performance Fortran (HPF) and good performance is attained for computations carried out on the Plasma Simulator NEC SX-7/160M5 at the National Institute for Fusion Science (NIFS). The code is applied to investigate the blob transport in magnetized plasma.

1 Introduction

Recent developments of parallel supercomputers enable researchers to perform 3D particle-in-cell (PIC) plasma simulations. In general, 3D PIC codes can be used for plasma phenomena related to fusion, laser plasma interaction, space technologies and processing. We have been developing a 3D PIC simulation code to investigate the dynamic plasma behavior in a Scrape of Layer (SOL) plasma. Because of the large computer resource required, it is necessary to use the parallel supercomputers efficiently.

In this article, we briefly describe the parallelization of our 3D PIC code using High Performance Fortran (HPF) [1] in Sect. 2 and present some recent results of a 3D bounded PIC simulation including blob transport in SOL plasma in Sect. 3.

Seiji Ishiguro
National Institute for Fusion Science, National Institute of Natural Science and Graduate University for Advanced Studies (SOKENDAI), Toki Gifu 509-5292, Japan,
e-mail: ishiguro.seiji@nifs.ac.jp

2 Parallelization of 3D Particle-in-cell Code using High Performance Fortran

In this section, we briefly describe the parallelization of 3D PIC code using HPF which is a distributed parallel processing language for Fortran. Programmers can parallelize a program by specifying the distribution of arrays and adding some directives. The Plasma Simulator NEC SX7/160M5 supercomputer system at National Institute for Fusion Science has five nodes, each node having 32 Processor Elements (PE). In order to use this supercomputer efficiently, a distributed memory parallelization across the nodes as well as shared memory parallelization within the nodes are necessary.

The PIC simulation code [2] consists basically of three parts: computing field quantities by solving Maxwell's equations, advancing velocities and positions of particles by solving a momentum equation, and determining the charge density and current density by depositing particle information to spatial grids. A typical simulation cycle of an electrostatic PIC simulation is shown in Fig. 1.

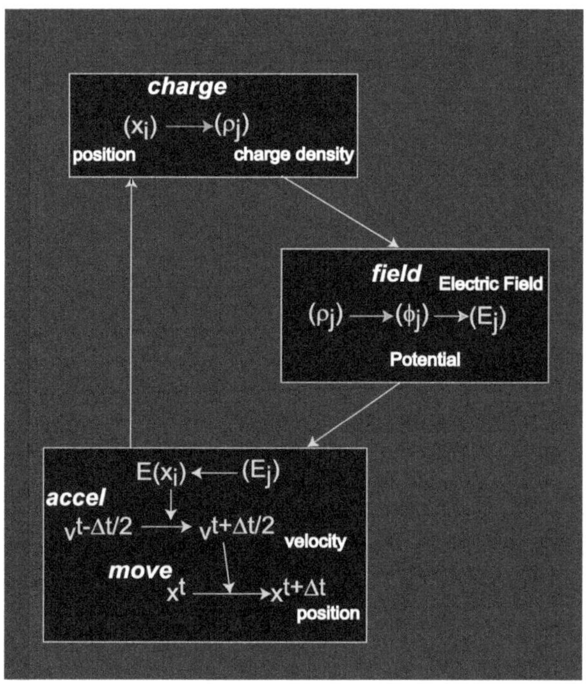

Fig. 1 Simulation cycle of an electrostatic PIC code

The PIC code uses much CPU time to calculate the deposition of particle positions onto grids and for the particle accelerations, while the CPU time spent for solving Maxwell's equations is quite small. Thus, the parts related to particle manipulation need to be parallelized efficiently. To use HPF for the distributed memory parallelization, we set up two dimensional arrays for particle velocities and positions and use a block distribution in the second index which corresponds to processor number. The shared memory parallelization makes use of the automatic parallelization of sxf90 compiler. The parts solving Maxwell's equations are only shared memory parallelized.

The parallelization is used to calculate a 3D electrostatic PIC simulation with periodic boundary conditions. Figure 2 shows the performance for different numbers of PEs for the case of a $128 \times 128 \times 512$ spatial grid and 1073741824 particles. The computation speed increases nearly linearly with the number of processors confirming the efficiency of the parallelization. The parallelized code has also been applied to relativistic electromagnetic PIC simulations and good performance was achieved [6, 3].

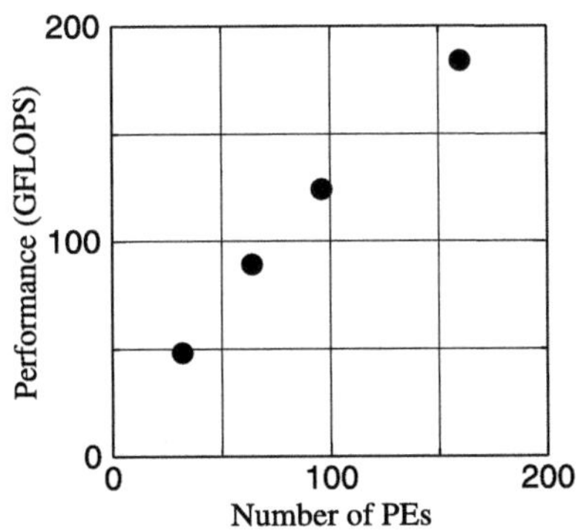

Fig. 2 Performance of the parallel PIC code for a 3D electrostatic PIC simulation

3 PIC Simulation of Blob Transport

As an example of large scale 3D PIC simulation, the blob transport in plasma is investigated. Recently, it was recognized that plasma is transported across magnetic field lines as long-living macroscopic coherent structures, so-called blobs, in the scrape-off layer of magnetic confinement fusion devices. Some theoretical and numerical studies present formulations of their dynamics of stability [7, 5]. In order to reveal more details about the microscopic physics related to blob dynamics, particle absorbing boundaries were integrated into the PIC code [4].

The configuration of our 3D PIC simulation is the following: an external magnetic field is pointing into the z-direction. Particle absorbing boundaries corresponding to diverter plates are placed at both ends of z-axis. A particle absorbing plate corresponding to the first wall is also placed at one end of the x-axis, a particle reflecting plane at the other one. Particle impinging onto the absorbing boundaries are removed from the system. In y-direction, periodic boundary conditions are applied.

The simulation parameters employed are the following: the system size $L_x \times L_y \times L_z$ is $256 \times 128 \times 2048$ and a $256 \times 128 \times 2048$ spatial grid system is used. The ion to electron mass ratio m_i/m_e is fixed at 100 and the ion to electron temperature ratio T_i/T_e is 0.25. The ratio of the electron cyclotron to electron plasma frequency ω_{ce}/ω_{pe} equals to 2.5, and the ratio of the ion cyclotron frequency to electron plasma frequency ω_{ci}/ω_{pe} is 0.025. The time step width is 0.05 where the time is normalized by ω_{pe}^{-1}. A density hump and grad-B in x-direction are employed. The magnetic field along the z-axis is given by

$$B_z(x,y,z) = B_0 \frac{512}{768 - x}$$

and the initial density profile by

$$n(x,y,z) \propto \left[\left(\frac{\tanh(x-128)}{32} + 1 \right) + 4\exp\left(- \left(\frac{x-192}{32} \right)^2 \right) \right] \times \left(4\exp\left(- \left(\frac{y-64}{32} \right)^2 \right) + 1 \right)$$

It is noted that averaged plasma density of the whole system is $n_0/2$.

Figure 3 shows several potential profiles, each one averaged over twice the plasma period. Here, the potential ϕ is normalized by e/T_e, where e is the electronic charge. The potential profile is almost symmetrical about the $z = 1024$ plane. On the left hand side, the potential is almost constant from the center of the system to the region at $z \approx 50$ in z-direction. It gradually decreases to $z \approx 5$ and then strongly decreases towards the boundary. This indicates an ion sheath is created in the vicinity of the boundary. The potential difference between the center of plasma region and the boundary is about 2, which is almost equal to the value of simple estimation $\phi = 0.5\ln(m_i/m_e) \approx 2.3$. The position at which the potential starts to decrease is $x \approx 50$. This is almost equal to $x = c_s t = 50$, where c_s is the ion acoustic speed. It is also clearly observed that a potential structure in the y-direction is present. The potential difference is approximately 0.5. The ions move in the positive y-direction and electrons move in the negative y-direction due to grad-B drift. The positive edge of the density hump is at $y \approx 100$, and thereby positively charged and the negative edge of the density hump at $y \approx 30$ is negatively charged. As a result, the potential difference in the y-direction is self-consistently created. This self-consistent electric field pointing into the negative y-direction plays a role in pushing the density hump in negative x-direction due to the $E \times B$ drift. Figure 4 shows the time evolution of ion density profile in the $x-y$ plane at $z = 1024$. The density hump moves in the negative x-direction at the speed $v \approx 0.1c_s$. At the same time, a small scale structure gradually appears in the center of the density hump. The mechanism of formation of this small scale structure is under investigation.

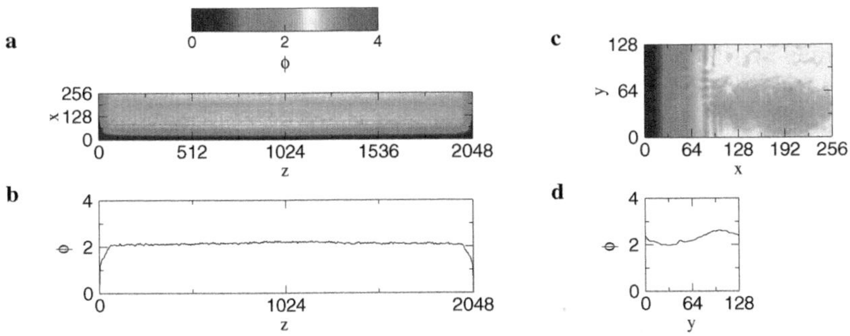

Fig. 3 Potential profiles in x-z plane at $y = 64$ **a** in the z-direction at $y = 64$ and $x = 192$ **b** in x-y plane at $z = 1024$ **c** and in the y-direction at $x = 192$, $z = 1024$ **d** at $t = 500$

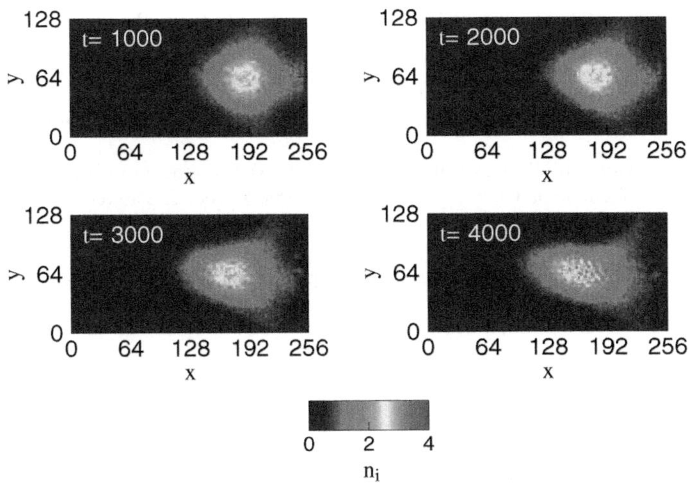

Fig. 4 Time evolution of ion density profile in the $x-y$ plane at $z = 1024$

4 Conclusion

We have developed a 3D PIC simulation code which is now parallelized using HPF and the automatic parallelization features provided by the sxf90 compiler. It is shown that the parallel version of the code performs well on the SX7/160M5 system at National Institute for Fusion Science / National Institute of Natural Science. A simulation demonstrates the creation of a self-consistent electric field which causes the transport of a blob across a magnetic field without any significant assumptions. The efficient use of HPC allows 3D PIC simulation which include very detailed physics compared to previous model analyses.

Acknowledgements This work is carried out with support and under the auspices of the NIFS Collaborative Research Program (NIFS06KTAT010). Computations have been performed on the Plasma Simulator (NEC SX-7/160M5) at NIFS.

References

1. High Performance Fortran Language Specification Version 2.0. High Performance Fortran Forum, Rice University, Houston, Texas (1997)
2. Birdsall, C.K., Langdon, A.B.: Plasma Physics via Computer Simulation. Adam Hilger, Bristol, Philadelphia and New York (1991)
3. Hasegawa, H., Ishiguro, S., Okamoto, M.: Distributed Parallelization of Exact Charge Conservation Particle Simulation Code by High Performance Fortran. In: J. Labatra, K. Joe, T. Sato (eds.) High-Performance Computing, *Lecture Notes in Computer Science, LNCS*, vol. 4759, 358. Springer, Heidelberg (2008)
4. Ishiguro, S. and Hasegawa, H.: Three-dimensional particle-in-cell simulation of scrape-off-layer plasmas. J. Plasma Physics **72**, 1233–1237 (2006)
5. Krasheninnikov, S. I.: On scrape off layer plasma transport, Physics Letters A **283**, 368–370 (2001)
6. Ohtani, H., Ishiguro, S., Horiuchi, R., Hayashi, Y., Horiuchi, N.: Development of Electromagnetic Particle Simulation Code in an Open System. In: J. Labatra, K. Joe, T. Sato (eds.) High-Performance Computing, *Lecture Notes in Computer Science, LNCS*, vol. 4759, 329, Springer, Heidelberg (2008)
7. Umansky, M., Krasheninnikov, S.I., LaBombard, B., Terry, J.L.: Comments on particle and energy balance in the edge plasma of Alcator C-Mod. Phys. Plasmas **5**, 3373–3376 (1998)

Multi-scale Modeling of Crack Propagation

Mitsuhiro Itakura, Hideo Kaburaki, Masatake Yamaguchi, and Ken-ichi Ebihara

Abstract Structural materials inside a nuclear reactor are subjected to contact with high-temperature coolant water, and after several decades of operation, cracks begin to grow in these materials which are otherwise resistant to cracks and corrosion. The mechanism of this phenomenon, referred to as "stress corrosion cracking", still remains unclear. In this paper, we present our attempt to identify the mechanism of stress corrosion cracking using numerical simulations, wherein we attempt to validate a hypothetical oxygen embrittlement mechanism. This mechanism involves diffusion of oxygen and embrittlement caused by the oxygen, and its simulation requires integration of oxygen diffusion and crack growth models. This paper presents the basic framework of such an integrated model and some preliminary simulation results, and shows that the integrated model captures the basic properties of stress corrosion cracking such as the rapid increase in the growth rate at the critical stress and its gradual increase above the critical stress.

1 Introduction

Inside a nuclear reactor, metallic structural components are subjected to neutron irradiation and contact with high-temperature coolant water. After several decades of operation, cracks can appear in components which are otherwise resistant to cracks and corrosion. This phenomenon is referred to as "Stress Corrosion Cracking" (SCC), since it occurs when the material is subjected to both tensile stress and a corrosive environment [5]. In SCC, cracks begin to grow at a weaker stress compared to a normal environment, making safety assessments of structural materials more subtle. Although SCC is known to occur for variety of combinations of materi-

Mitsuhiro Itakura · Hideo Kaburaki · Masatake Yamaguchi · Ken-ichi Ebihara
Center for Computational Science & e-System, Japan Atomic Energy Agency, Taito-ku, Higashiueno 6-9-3, Tokyo 110-0015, Japan,
e-mail: {itakura.mitsuhiro, kaburaki.hideo, yamaguchi.masatake, ebihara.kenichi}@jaea.go.jp

als and environments, the mechanism of interaction between stress and the corrosive environment is still not fully understood.

Intensive experimental studies have been carried out to measure SCC growth rates, and long-term SCC growth in an operational reactor is extrapolated from these results. Unfortunately, the lack of knowledge of the mechanism of SCC causes an uncertainty in the extrapolation formula. On the other hand, if the mechanism of SCC can be identified, we can provide a theoretical foundation for these extrapolations and hopefully reduce this uncertainty. This topic has therefore been intensively studied, and various hypothetical mechanisms have been proposed. Among them are oxide film rupture at the crack tip, loss of corrosion resistance at grain boundaries by irradiation-induced chrome depression, hydrogen or oxygen embrittlement, and localized plasticity (see [7] and references therein for recent studies). Currently there is no consensus as to which mechanism is the most plausible one, however, recent trends seem to indicate that the oxygen embrittlement scenario is gaining increasing support [2]. To validate the oxygen embrittlement mechanism, we have constructed a multi-scale simulation model of SCC based on it and have compared the results with experiments.

This paper is organized as follows. First, the basic framework of the multi-scale simulation model is presented. Then, details of the numerical simulation and some preliminary results are shown. Finally, the conclusions and plans for future studies are presented.

2 Multi-scale Model of SCC

2.1 Oxygen Embrittlement Mechanism

The oxygen embrittlement scenario of SCC, schematically shown in Fig. 1, has been proposed by several authors based on experimental observations [2]. First, oxygen atoms contained in the environmental water penetrate into the boundaries of polycrystalline grains (grain boundaries), since there is more interstitial atomic space at grain boundaries than in the bulk regions into which oxygen atoms can diffuse. In addition, the tensile stress acting on the grain boundaries increases the inter-atomic distance at grain boundaries and makes more room for oxygen atoms. When sufficient quantity of oxygen atoms diffuses into the grain boundaries, the inter-atomic cohesive force, which "glues" crystal grains on both sides of the grain boundary, becomes weaker, making the critical stress for crack growth smaller, too. When this critical stress becomes smaller than the external tensile stress, the crack proceeds to this weakened region. Water in the environment fills the cracked region, and oxygen atoms can diffuse deeper into the material and weaken subsequent regions of grain boundaries. By repetition of this process, the crack grows into the material. This hypothetical mechanism is consistent with the typical features of SCC observed in nuclear reactor structural materials, such as

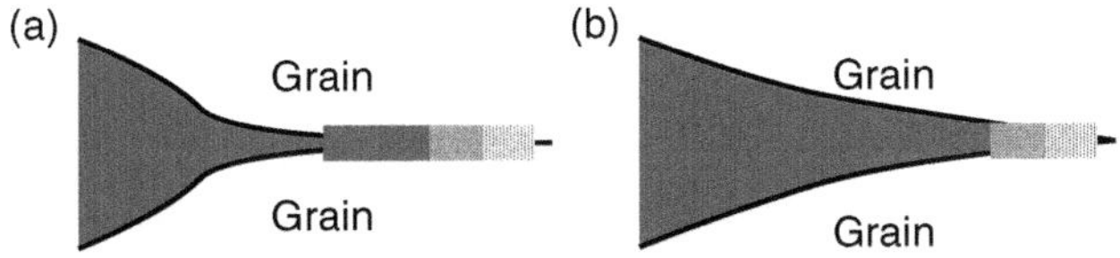

Fig. 1 Schematic depiction of the oxygen embrittlement mechanism of SCC. **a** first, oxygen atoms diffuse into the grain boundary ahead of the crack tip (indicated by the shaded region) and embrittle the crack tip region **b** the crack proceeds to the embrittled region, enabling oxygen atoms to diffuse deeper into the material

its lower critical stress and slower growth rates compared to a normal environment, since the controlling factor of crack growth rate in this scenario is oxygen diffusion which is very slow, and the oxygen lowers the critical stress for crack growth.

Our objective was to carry out crack growth simulation based on this mechanism and compare the results with experimental data to validate the mechanism. In this work, we followed the multi-scale modeling scheme developed by Serevrinsky and Ortiz [3]. Basically, the model consists of two parts: diffusion of oxygen and mechanics of crack growth. The diffusion of oxygen along grain boundaries is simulated by solving a standard diffusion equation, and the effects of oxygen on crack growth is incorporated using the so called "cohesive element" whose response to stress depends on the oxygen density.

2.2 Mechanics Modeling of Crack Growth

The trickiest part of finite element calculation of crack growth is "on the fly" determination of the crack path and subsequent re-meshing. However, this is not a problem in SCC since the cracks follow pre-determined grain boundaries. The crack growth along the grain boundaries is modeled using cohesive elements, which have zero thickness and are placed at each grain boundary [3]. Basically, the cohesive elements "glue" two crystal grains on both sides, and are characterized by cohesive force between crystal grains as a function of the distance between the two crystal grains. When the distance is as short as the range of the inter-atomic force, the cohesive force is approximately proportional to the distance, while at longer distances, the force gradually drops to zero and remains so thereafter. This cohesive force can be directly calculated by quantum mechanical calculations. Yamaguchi et al. have carried out quantum mechanical calculations of cohesive force of nickel grain boundaries, with and without sulfur atoms, and found that sulfur atoms drastically reduce the cohesive force [6]. We expected that oxygen has a similar effect on grain boundaries of iron, which is the matrix material of the stainless steels used for structural components of reactors. Currently, quantum calculations of the effect of oxygen on the grain boundary cohesive force are in progress. In the present study, we used a tentative cohesive force of the iron grain boundary with oxygen, shown in Fig. 2. We assume that oxygen reduces the cohesive force by a factor of one tenth.

The rest of the material is modeled by a standard finite-element method of elasticity. The shapes of the crystal grains are generated by random Voronoi tessellation. The typical size of the grains is about 50 μm, and we used 1,000 to 12,000 grains in the simulations. Figure 3 shows the configuration of 12,000 grains. For efficient calculation, an initial crack, which is perpendicular to the external applied stress direction and has length of one fourth of the system size, is introduced at one end of the simulation system.

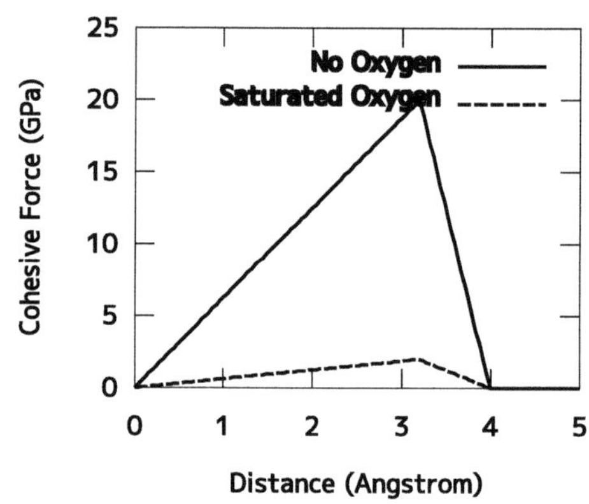

Fig. 2 Tentative cohesive force of an iron grain boundary without oxygen and with saturated oxygen atoms (dotted and solid lines, respectively) used in the present work

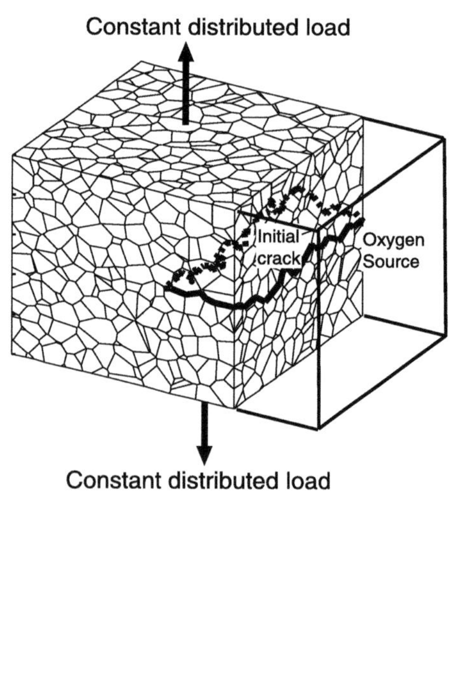

Fig. 3 Configuration of 12,000 grains generated by random Voronoi tessellation, with the initial crack perpendicular to the external applied stress

2.3 Oxygen Diffusion Modeling

The diffusion of oxygen was simulated by solving the standard diffusion equation. In the present study, we only considered grain boundary diffusion and omitted bulk diffusion inside grains, which is negligible. Thus, the diffusion equation is solved on a network of random-shaped polygons, connected to each other at their edges. The diffusion equation becomes as follows:

$$dC(r)/dt = -k(r)\nabla^2 C(r)$$

where $C(r)$ denotes the oxygen density at position r on some grain boundary, $k(r)$ denotes the diffusion constant at r, and the operator ∇ is defined in two-dimensional space. Quantum mechanical calculations show that oxygen atoms are accommodated at vacancy sites of grain boundaries, and when all of these sites are occupied by oxygen, no more oxygen can be accommodated. We normalized the oxygen density by this saturated density. The diffusion constant $k(r)$ is related to the energy barrier between two vacancy sites on a grain boundary, and it can be directly evaluated from quantum mechanical calculations. It is possible that this energy barrier depends on the Σ-value of grain boundary and tensile stress acting on it. This kind of dependence can also be evaluated by quantum mechanical calculations.

The mechanics model and the diffusion model were coupled together in such a way that the diffusion of oxygen changed the strength of the cohesive force, and redistribution of stress changed the diffusion constant. In addition, when separation between two grains was sufficiently wide, the interim space between them can be regarded as an external environment and the boundary condition for the diffusion equation needs to be updated accordingly. In the simulation, the mechanics and diffusion calculation were carried out alternately. Since the time scale of the mechanics is orders of magnitude shorter than that of diffusion, the time step of the simulation is taken to be that of diffusion, and mechanics part is equilibrated at each step for a fixed density of oxygen.

3 Simulation Details and Results

To implement simulation of the multi-scale model presented in the previous section, a finite element mesh needs to be generated which is compatible with the shape of the crystal grains. Such a mesh can be created by first generating a triangular mesh on each grain boundary surface and then generating a tetrahedral mesh inside each grain compatible with the surface mesh. To evaluate the stress distribution precisely, a fine mesh must be used at or around grain boundaries where the crack might proceed. In the present paper, however, we used a simpler mesh generation scheme described in [1] for preliminary simulations. For the diffusion equation, we applied a finite-volume discretization scheme to the triangular surface mesh, and time development was solved by a second-order explicit Euler scheme. For simplicity, we used a tentative value for the diffusion parameter k, which is assumed to be independent of the applied stress and type of grain boundary.

In the case of 12,000 grains, the degree of freedom for the mechanics part is 1,500,000 and the number of non-zero elements in the elasticity matrix is 28,000,000, while the diffusion problem is basically two-dimensional and the degree of freedom is orders of magnitude fewer than the mechanics part. Therefore, most of the CPU time is consumed in the mechanics part, especially in matrix-vector product calculations of the solver routine for linear elastic equations. By using a vectorized solver routine on the SX-7 vector processor, we attained an overall vectorization ratio of over 95%.

Currently, we are experimenting with preliminary simulation settings to estimate the optimum simulation parameters such as the mesh size and the number of grains, before implementing a full-scale simulation. Here, we present some results of the preliminary simulation, in which tentative values for the cohesive force and diffusion constant are used. Figure 4 shows the oxygen density on grain boundaries on and ahead of a crack, in a system composed of 1,000 grains. Note that the displacement due to the tensile stress is exaggerated 100 times in the figure. In Fig. 4, one can see the diffusion of oxygen from the environment into the grain boundaries ahead of the crack. The red color indicates saturated oxygen density $C = 1.0$. A cracked grain boundary is regarded as being open to the environment, and the boundary condition of oxygen diffusion is modified so that the density on a newly created crack surface is fixed at 1. Without oxygen, the applied stress in this simulation is not strong enough for crack growth, while for grain boundaries with sufficient oxygen, it is sufficient for crack growth. Therefore, the crack does not grow at first, however, after the diffusion of oxygen, it begins to grow. When the external stress is stronger, the oxygen density required for crack growth becomes smaller and the crack growth rate becomes faster. On the other hand, when the external stress is not

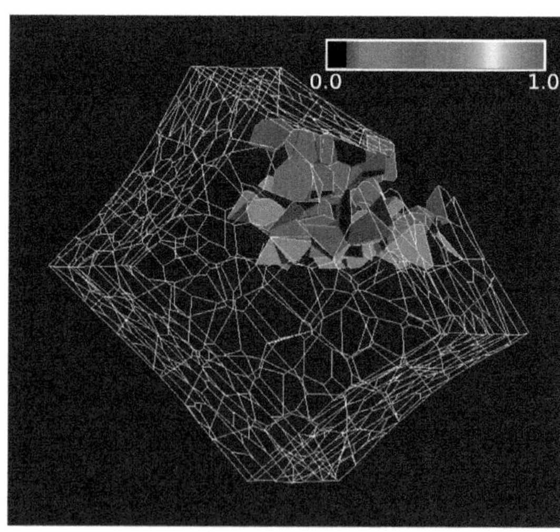

Fig. 4 Three dimensional visualization of oxygen density (indicated by the colors) on grain boundaries ahead of a crack. The density 1.0 (red) corresponds to saturated density.

strong enough even for saturated oxygen density, the crack does not grow. These properties are a unique result of integration of the diffusion and crack growth models. Figure 5 shows the crack growth rate plotted as a function of the tensile stress applied to the system, together with experimental data taken from [4]. It should be noted that we have adjusted the diffusion constant so that the absolute values of the growth rate coincide with the experimental ones, and the main purpose of this preliminary simulation is to verify *qualitative* agreement of basic properties of the growth rate with experiments. From this viewpoint, one can see that the basic properties of the simulation data, such as the rapid increase in the growth rate at the critical stress and its gradual increase above the critical stress, agree well with experiments.

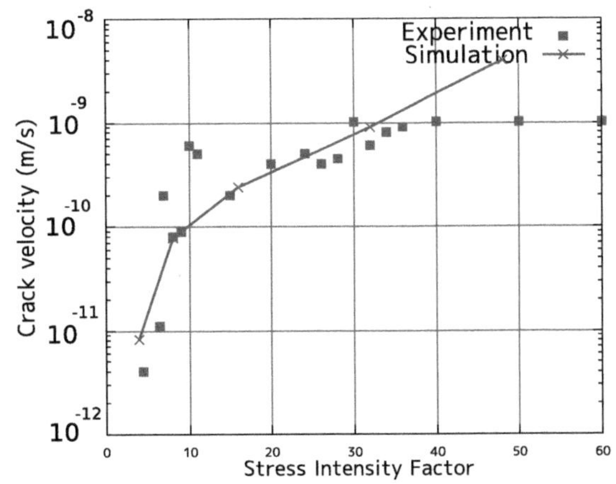

Fig. 5 Crack growth rate obtained by preliminary simulation, plotted as a function of the external stress (solid line), together with experimental data (solid squares) from [4]. See text for details.

4 Conclusions

To validate the oxygen embrittlement mechanism of SCC, we have developed a multi-scale model of SCC which integrates diffusion simulation of oxygen and crack growth simulation. The developed model is capable of capturing the interaction between diffusion and crack growth, and the simulation results agree well with the basic properties of SCC crack growth rates observed in experiments, such as a rapid increase in the growth rate at critical stress and its gradual increase above the critical stress. Simulations were optimized for vector processors, and a vectorization ratio of over 95% was attained on the SX-7 vector processor.

For the validation of the oxygen embrittlement mechanism, a quantitative comparison between the simulation results and experiments is necessary, and it requires parameters evaluated by quantum mechanical calculations and a finer finite element mesh. These quantum calculations and the development of a finer mesh are currently in progress.

Acknowledgements The authors would like to thank Prof. M. Ortiz (Caltech), Dr. J. Rimoli (Caltech) and Dr. M. Kamata (INSS) for essential discussions.

References

1. Itakura, M., Kaburaki, H., Arakawa, C.: Branching mechanism of intergranular crack propagation in three dimensions, Phys. Rev. E **71**, 055102 (2005)
2. Lu, Y.H., Peng, Q.J., Sato, T., Shoji, T.: An ATEM study of oxidation behavior of SCC crack tips in 304L stainless steel in high temperature oxygenated water, Journal of Nuclear Materials **347**, 52–68 (2005)
3. Serebrinsky, S., Carter, E. A., Ortiz, M.: A quantum-mechanically informed continuum model of hydrogen embrittlement, Journal of the Mechanics and Physics of Solids **52**, 2403–2430 (2004)
4. Shunji, S., Totsuka, N., Kamaya, M., Nakajima, N.: Research on Crack Growth Rate of Primary Water Stress Corrosion Cracking of MA600 Alloy, INSS Journal **7** (2000)
5. Staehle, R.W., et al. (eds.): Proceedings of the Conference on Stress Corrosion Cracking and Hydrogen Embrittlement of Iron Base Alloys, NACE, Houston (1977)
6. Yamaguchi, M., Shiga, M., Kaburaki, H.: Grain Boundary Decohesion by Impurity Segregation in a Nickel-Sulfur System Science **307**, 393 (2005)
7. Young, G., Wilkening, W., Morton, D., Richey, E., Lewis, N.: The Mechanism and Modeling of Intergranular Stress Corrosion Cracking of Nickel-Chromium-Iron Alloys Exposed to high Purity Water, Environmental Degradation of Materials in Nuclear Systems **XII** (2007). CD-ROM. http://www.osti.gov/energycitations/product.biblio.jsp?osti_id=850543

V
Climate Modeling, Hydro- and Aerodynamics

The Climate Model ECHAM5 on NEC SX-8

Stefan Borowski

Abstract Climate simulation codes like ECHAM5 consider the dynamics of the Earth system over a long period of time. The huge number of time steps to be propagated restricts the spatial discretization to comparably coarse grid resolutions. This requires scalable parallelizations based on the lowly resolved spatial domain which can cope with the computational effort in the non-parallelizable time domain. The paper on hand gives an overview of the optimization of ECHAM5 on the NEC SX-8 with respect to performance and scalability. Several single CPU and scalability optimization are described. The outcome of the optimizations is demonstrated in detailed performance measurements.

1 Introduction

ECHAM5 is the 5th generation of the atmospheric circulation model developed by the Max-Planck-Institute for Meteorology at Hamburg [1, 2]. It describes the dynamics of Earth's atmosphere including physical, chemical, biological processes and in particular contributions by human behavior. Simulations with ECHAM5 are part of Earth system research strategically aiming at prediction of climate dynamics.

The numerical implementation of ECHAM5 is characterized as follows. Time is discretized by semi-implicit leapfrog time differencing. Space is discretized on a structured three-dimensional grid: latitudes and longitudes in horizontal directions and layers in vertical direction. Differential operators in horizontal directions are applied in spectral space (e.g. horizontal diffusion). For this purpose the representation is changed from spatial to spectral space by Legendre transforms and Fast Fourier Transforms (FFTs). Preceding the transformations, data transpositions

Stefan Borowski
NEC Deutschland GmbH, Hessbruehlstr. 21b, 70565 Stuttgart, Germany,
e-mail: sborowski@hpce.nec.com

move the data to be transformed to the leading dimension. Differential operators in vertical direction are applied in spatial space by Finite Differencing.

Most computations are performed in spatial grid representation by collapsed and blocked loops over longitudes and latitudes. The runtime parameter NPROMA sets the block size and thus the length of the innermost loop. The resulting number of blocks NGPBLKS gives the length of the outermost stripped loop.

A hybrid parallelization combines MPI and OpenMP. The MPI parallelization is based on a two-dimensional NPROCA×NPROCB domain decomposition. The runtime parameters NPROCA and NPROCB determine the number of domains along a longitude (north-south direction) and a latitude (east-west direction), respectively. To compensate load imbalances each domain in north-south direction covers two symmetric areas: one on the northern and the other on the southern hemisphere. The introduced data transpositions gathering data for the transforms result in an all-to-all communication pattern. The OpenMP parallelization relies on loop based parallel regions. Within the blocked loop scheme the outer stripped loop over blocks is parallelized.

Three resolutions are considered for evaluation. T63L31 is a typical production resolution of size $N_S = 192 \times 96 \times 31$ [1]. The larger resolution T255L60 of size $N_S = 768 \times 384 \times 60$ gives more detailed results for selected studies. T511L91 of size $N_S = 1536 \times 768 \times 91$ is a huge resolution to inquire the feasibility of finer grid resolutions for future studies.

2 Runtime Analysis

Profiles of ECHAM5 runs show that the computing time is evenly distributed among various procedures. The most expensive procedures are:

- Cloud physics on blocked grid representation
- Legendre transforms (matrix-vector multiplications) and FFTs
- Field transport
- Data transpositions (pack, communicate, unpack)
- Vertical pressure grid mapping
- Intrinsic power function

While single CPU performance is reasonable on a small number of domains, scalability with increasing number of domains is unsatisfactory. The analysis of this lack of scalability revealed a critical point in climate simulations. Climate researchers are interested in simulating long periods of time (100 – 1000 years) which requires a huge number of time steps $N_T = 10^6 - 10^7$. For an acceptable time to solution such long time propagation allows only a coarse resolution of Earth resulting in problem

[1] Resolution notation Tn_{max}Ln_{lay} of size $N_S = n_{lon} \times n_{lat} \times n_{lay}$: triangular truncation of spherical harmonics at meridional wave number n_{max} corresponding to n_{lat} latitudes and n_{lon} longitudes in horizontal directions, n_{lay} layers in vertical direction.

sizes $N_S = 10^5 - 10^6$. In fact, the unusual relation $N_T > N_S$ causes a scalability problem: Climate simulations require High Performance Computing (HPC) to cope with the large computational effort in the non-parallelizable time domain. However, the parallelization utilizing the HPC resources can only refer to data objects of small size in the lowly resolved spatial domain.

3 Optimizations

To overcome this scalability problem, the following optimizations address two points: improvement of single CPU performance to provide a basis for scalability and improvement of scalability to sustain the reached performance per CPU up to large CPU counts.

3.1 Single CPU Optimizations

Two main issues limiting the single CPU performance are identified. On one hand, the average vector length is not satisfying for several procedures due to small length of innermost loops. On the other hand, most procedures are limited by memory access due to low density of data processing. Therefore, the applied optimizations enlarge average vector length (indicated by VL) and reduce memory access (indicated by MA) [3]:

- Combine matrix-vector multiplications (VL, MA):
- Unroll outer loops into vectorized loops (MA)
- Fuse and compact vectorized loops (MA)
- Expand loops with known iteration count within vectorized loops (MA)
- Manually expand routines inline to enable loop collapsing and unrolling (VL, MA)
- Manually collapse loop nests involving assumed shape arrays (VL)
- Strip mine and introduce vector registers (MA)
- Simplify or resolve if-clauses depending on inner loop index (MA)
- Move independent if-clauses outside (MA).

Moreover, the vertical pressure grid mapping is completely reworked. Finally, the intrinsic power function is replaced by a more efficient computation (compiler switch `-Wf,-pvctl chgpwr`).

The larger average vector length and the reduced memory access improve the continuous data supply to the vector pipelines which directly implies a better sustained performance on the vector processor (see Sec. 4.1).

3.2 Scalability Optimizations

The total number of iterations within one time step corresponds to the small problem size resulting from low spatial resolution. Here, a better scalability can only be achieved by evenly distributing these iterations among all levels of parallelism:

- Massively Parallel Processing (MPP): MPI parallelization
- Symmetric MultiProcessing (SMP): OpenMP parallelization
- Instruction Level Parallelism (ILP): Vectorization

Load balancing between these three levels is ensured by appropriate domain decomposition and blocking (see Sec. 4.2). The improvement of vectorization is discussed in the previous Sec. 3.1.

The implemented optimizations of the OpenMP parallelization not only improve load balancing (indicated by LB) but also reduce parallel overhead (indicated by PO) [4]:

- Simplify or resolve if-clauses depending on outer loop index (LB)
- Introduce dynamic worksharing (LB)
- Merge and extend loop based parallel regions (PO)
- Manually collapse parallel loops with other loops to keep all threads busy (LB)
- Save redundant synchronizations between workshare constructs (PO)

In MPI communication, considerable time is spent on packing and unpacking data to be communicated. To save redundant internal copying into Global Memory all communication buffers used for transpose, gather and reduce operations are directly allocated in Global Memory.

4 Performance Results

4.1 Single CPU Performance

Figure 1 shows the performance per CPU of ECHAM5 for optimized (black) and original (gray) version on T63L31 and T255L60. The two resolutions are computed on 4 as well as 8 CPUs with 1 MPI process per CPU. A decomposition in only latitudes is used (`NPROCB=1`). The block size `NPROMA` is set to 4608 and 2304 on 4 and 8 CPUs, respectively.

An average performance improvement of 31% is achieved by the single CPU optimizations. However, the pure MPI runs also illustrate the scalability issue: The fall of single CPU performance by both decreasing problem size (cf. T255L60 and T63L31) and increasing number of used CPUs (cf. 4 and 8 CPUs) is significant. The next section shows how this issue is addressed by the optimized hybrid parallelization.

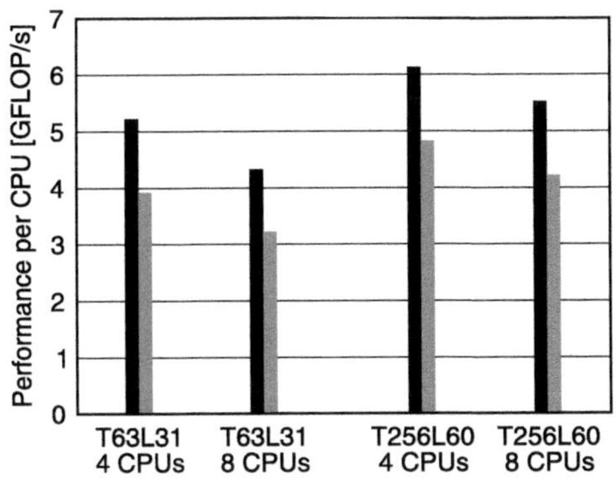

Fig. 1 Single CPU performance of ECHAM5 on T63L31 and T255L60 in pure MPI runs on 4 and 8 CPUs for optimized (black) and original (gray) version.

4.2 Scalability

In order to satisfy all levels of parallelism (MPP, SMP, ILP) the domain decomposition as well as the blocking scheme should be even. The primary grid representation in spatial space of size $N_S = n_{lon} \times n_{lat} \times n_{lay}$ is to be decomposed in NPROCA×NPROCB domains on unity sphere. As distributing longitudes results in shorter average vector length for the Legendre transforms, a decomposition in only latitudes is employed (NPROCB=1). By choosing NPROCA as factor of n_{lat} all MPI processes work on equally sized domains. For the blocking scheme there are two requirements. To fully utilize the vector processor, NPROMA is chosen as multiple of the hardware vector length MAXVL=256. To equally employ all OpenMP threads, NGPBLKS is chosen as multiple of the number of threads. Complying with these constraints the choice of NPROCA, NGPBLKS and NPROMA fullfills the equation

$$n_{lon} \times n_{lat} = \mathrm{NPROCA} \times \mathrm{NGPBLKS} \times \mathrm{NPROMA}. \tag{1}$$

Figure 2–4 show the scaling behavoir of ECHAM5 for optimized (black) and original (gray) version on the three considered resolutions. The total performance and forecast days per day depending on the number of used CPUs are illustrated. The block size NPROMA is set to 768 on T63L31 and T255L60 whereas to 2304 on T511L91. One MPI process is run on each SX-8 node with according number of OpenMP threads. Generally, total performance is in proportion to forecast days per day. Thus, performance directly translates into reduction of time to solution. However, the improvement rate for forecast days per day is higher than that for total performance due to omitting redundant operations in the optimized version.

For T63L31 the forecast days per day rate is improved by 73% to 17500 on 3 SX-8 nodes. The discontinuity between 8 and 12 CPUs in the curve for the original version indicates memory congestion. In the 8 CPU run on a single node, all eight CPUs of the node compete for memory. In the 12 CPU run on two nodes, however, only six CPUs per node compete for memory. Therefore, the performance per CPU

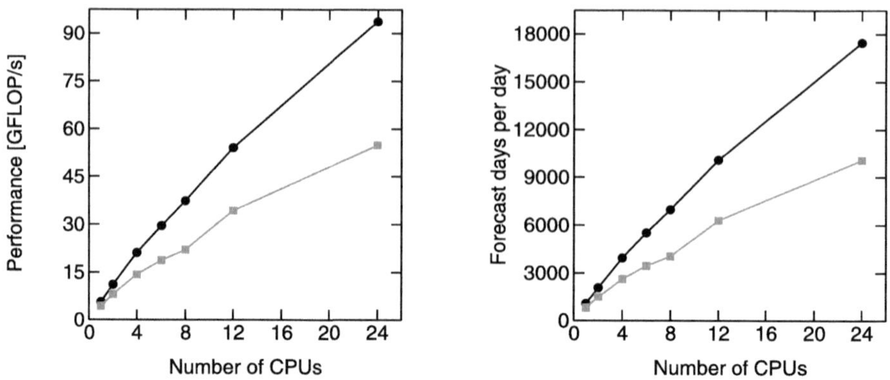

Fig. 2 Scaling behavior of ECHAM5 on T63L31: total performance and forecast days per day for optimized (black) and original (gray) version.

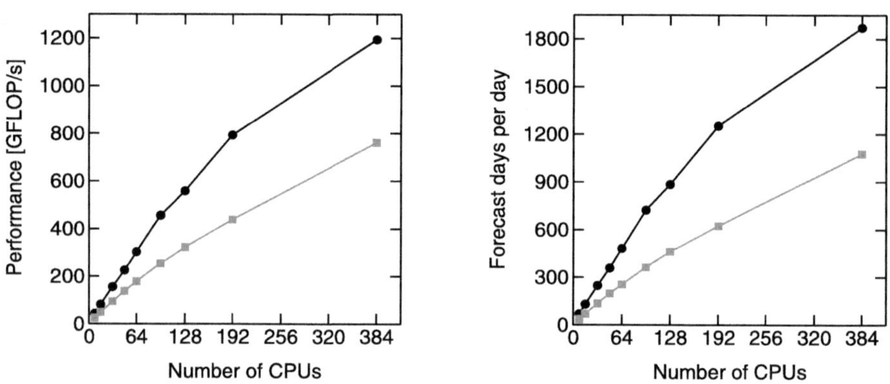

Fig. 3 Scaling behavior of ECHAM5 on T255L60: total performance and forecast days per day for optimized (black) and original (gray) version.

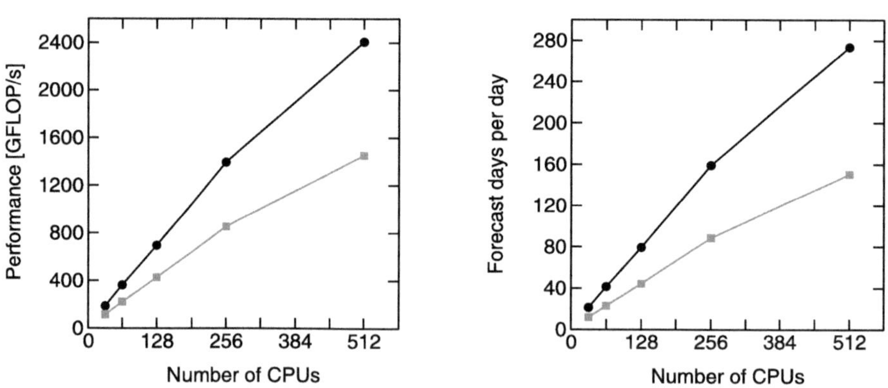

Fig. 4 Scaling behavior of ECHAM5 on T511L91: total performance and forecast days per day for optimized (black) and original (gray) version.

is higher in the 12 CPU run. In comparison to this finding the curve for the optimized version shows a continuous progression. Hence, memory congestion can be avoided by proper scheduling of memory access.

For T255L60 the forecast days per day rate is improved by 74% on 48 SX-8 nodes. Thereby, 1 TFLOP/s sustained performance is exceeded with a total performance of 1.2 TFLOP/s. However, the decrease of parallel efficiency is more significant than for the other two resolutions. This becomes understandable by regarding local problem size in relation to number of used CPUs: On 48 nodes each of the 384 CPUs works on 46080 grid points. In contrast, there are an order of magnitude less CPUs for T63L31 (23808 grid points per CPU on 24 CPUs) and an order of magnitude more grid points for T511L91 (209664 grid points per CPU on 512 CPUs).

For T511L91 the forecast days per day rate is improved by 81% on 64 SX-8 nodes. Thereby, the top total performance of 2.4 TFLOP/s is reached. With each of the 512 CPUs working on 209664 grid points the highest parallel efficiency is maintained. It amounts to 80% referring to the 4 node run as starting point.

5 Conclusions

The atmospheric climate model ECHAM5 was analyzed, optimized and evaluated. The found data processing is typical of climate simulations: a very long time propagation on low spatial resolutions. Therefore, the main computational effort is due to time propagation while the coarse grid resolution results in relatively small data objects. As the time domain is not parallelizable, there is the challenging demand for an efficient parallelization referring to these small data objects.

As a prerequisite for the refinement of the existing hybrid parallelization, a single CPU performance improvement of 31% was achieved by inline expansion of procedures, optimization of if-clauses and several loop transformations. For the hybrid parallelization the main effort was put into the improvement of load balancing. Iterations within one time step were evenly distributed among all levels of parallelism: Massively Parallel Processing (MPP), Symmetric MultiProcessing (SMP) and Instruction Level Parallelism (ILP). Load imbalances and parallel overhead in SMP were reduced by rearranging if-clauses, merging parallel regions and introducing dynamic worksharing. On average an improvement of 76% was achieved for the largest CPU count. The top performance of 2.4 TFLOP/s was reached for T511L91 on 64 SX-8 nodes. These results demonstrate that an efficient parallelization masters the enormous computational effort by long time propagation and makes the climate simulation scale up to a large number of CPUs.

Acknowledgements We are grateful to Luis Kornblueh for interesting and fruitful discussions.

References

1. Special section: Climate models at the Max Planck Institute for Meteorology (MPI-M). Journal of Climate **19**, 3769–3987 (2006)
2. Kornblueh, L.: ECHAM5 - an atmospheric climate model and the extension to a coupled model. In: M. Resch, T. Bönisch, S. Tiyyagura, T. Furui, Y. Seo, W. Bez (eds.) High Performance Computing on Vector Systems. Proceedings of the High Performance Computing Center Stuttgart, March 2006, pp. 171–181. Springer, Berlin Heidelberg, Germany (2007)
3. NEC Corporation: SUPER-UX Performance Tuning Guide. Revision 1.2 (2007)
4. NEC Corporation: SUPER-UX Shared Memory Parallelization Guide. Revision 1.2 (2007)

A Large Spectrum of Free Oceanic Oscillations

Malte Müller

Abstract An ocean model has been developed to compute a large set of barotropic free oscillations with explicit consideration of dissipative terms and the full ocean loading and self-attraction effect. The Implicitly Restarted Arnoldi Method is utilized to determine these free oscillations. It is a highly efficient approach to solve large scale eigenvalue problem, in particular if the matrix entries are generally nonzero. The mean performance on the SX-8 supercomputer is up to 3.4 TFlops on 512 CPUs.

1 Introduction

Barotropic oceanic oscillations are existing on various timescales in the global ocean. It is of considerable importance to understand the oscillation behavior of the ocean, e.g. due to the significant interactions with the ocean circulation or to prevent aliasing errors by processing satellite data. The most prominent oscillations are the ocean tides. The strongest tides have periods in the semidiurnal and diurnal time range. Meanwhile, ocean models are computing the tides with very high accuracy due to the assimilation of satellite data. Further, these experiments depict model deficiencies represented through so called dynamical residues. It is very important to understand these dynamical residues in order to learn about the tides of the real ocean. As well, oceanic oscillations are atmospherically forced on timescales longer than two days. So far, it is an open questions whether these long periodic oscillations are resonantly forced oscillations or are in equilibrium with the driving force.

Malte Müller
Institute of Oceanography, Bundesstr. 53, 20146 Hamburg, Germany,
e-mail: malte.mueller@zmaw.de

2 State of Knowledge

In the last century approaches to determine the free oscillations of different kinds of schematic basins have been developed. However, in a rotating frame of reference these solutions are not available in a closed form. A detailed representation of the free oscillations of a rotating rectangular basin, including experimental studies, is given by [19]. Free oscillations of the rotating Earth fully and hemispherically covered by water, are obtained by [12]. These solutions are generalized through the consideration of the full loading and self-attraction (LSA) effect by the studies of [26]. The first numerical solutions of free oscillations for a frictionless global ocean with realistic topography on a rigid earth were obtained by [7, 18, 5]. [13] investigated the long period vorticity modes and the effect of bottom topography using the model from [18], but restricted the investigation to the Pacific Ocean. A set of free oscillations of a barotropic global ocean model with realistic topography and explicit consideration of dissipative terms was computed by [28]. There, the LSA-effect was introduced by parametrization and the full LSA-term was considered for only two modes. In barotropic ocean dynamics the secondary effect of ocean loading and self-attraction is known to be an essential part. It is often considered in a simplified manner, because the full LSA-term turns the dynamical equations into an integro-differential equation system that makes consideration of the full effect very time consuming in numerical models. However, a recent review of the LSA-effect recommends that most serious applications should use the full integral formulation [20]. This convolution integral is defined through the so called Green's function of loading and self-attraction (see Sect. 3). The Green's function are given in terms of spherical harmonics, weighted through the degree-dependent loading Love numbers k'_n and h'_n, which are computed by an earth model considering the features of elasticity and the radial density distribution of the earth [3]. The Green's function is written as

$$G(\lambda,\phi,\lambda',\phi') = \frac{1}{4\pi}\sum_{n=0}^{\infty}(1+k'_n-h'_n)\,\alpha_n\sum_{s=0}^{n}\overline{P}_{n,s}(\sin\phi)\,\overline{P}_{n,s}(\sin\phi')\cos(s(\lambda'-\lambda))\,. \tag{1}$$

An important characteristic of this Green's function is that its dependency on the four variables $(\lambda,\phi,\lambda',\phi')$ can be reduced to

$$G(a) = \frac{1}{4\pi}\sum_{n=0}^{\infty}(1+k'_n-h'_n)\alpha_n\overline{P}_n(\cos a). \tag{2}$$

on the angular distance a. The Green's function, determined by the loading Love-numbers of [4], is displayed in Fig. 1. Schematically, the LSA effect is shown in Fig. 2. There, at the position of the wave crest the additional mass deforms the ocean bottom in downward and at the wave minimum in upward direction. This

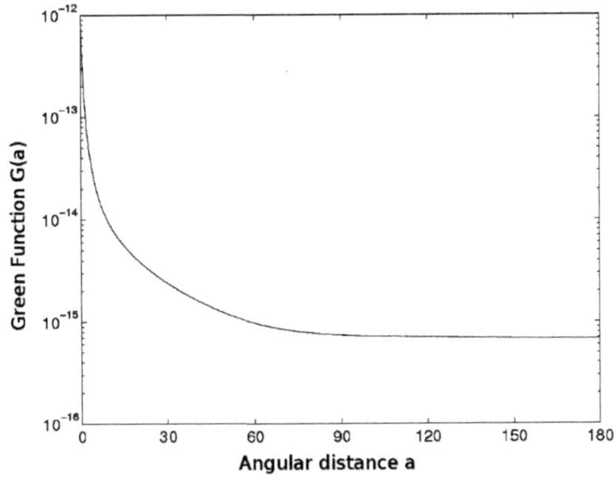

Fig. 1 The Green's function depending on the angular distance a between the two points (ϕ, λ) and (ϕ', λ'). It holds for the angular distance: $(\cos(a) = \sin\phi \sin\phi' + \cos\phi \cos\phi' \cos(\lambda - \lambda'))$, (Data from [4])

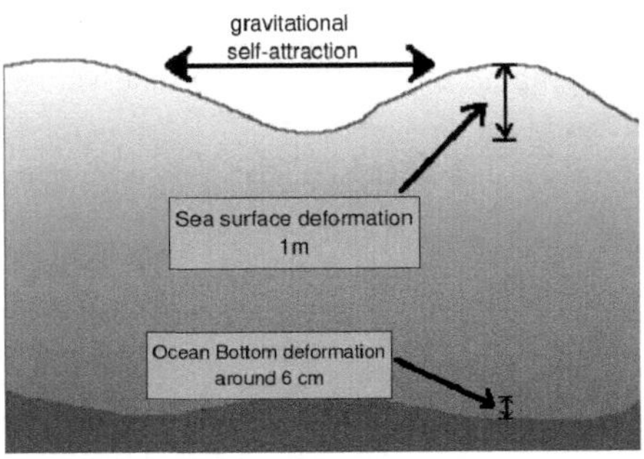

Fig. 2 Schematical figure explaining the loading and self-attraction effect: The mass load of the wave (amplitude one meter) deforms the ocean bottom by around 6 cm (loading effect). The two wave crests attract each other due to their mass (self-attraction effect).

effect is called the loading effect. If one assumes an amplitude of one meter for an ocean wave, the ocean bottom gets deformed by around 6 cm. The self-attraction effect is caused through the additional gravitation force caused by the mass of the wave crest.

3 Free Oceanic Oscillation Model with Consideration of LSA

To understand why ocean tides are as they are and to clarify the role of resonant long periodic oscillations it is substantial to know about the free oscillations of the global ocean. The determination of these free oscillations with explicit consideration of all relevant effects needs large computer resources and a highly efficient modeling approach. A unique model approach has been developed and for the first time free oscillations are determined with explicit consideration of frictional terms and the full LSA effect.

3.1 Theory

The barotropic free oscillations of the global ocean are defined through the linearized homogeneous shallow water equations (e.g. [25]).

$$\frac{\partial \mathbf{v}}{\partial t} + \mathbf{f} \times \mathbf{v} + \frac{r'}{D}\mathbf{v} + \mathbf{F} + g\nabla\zeta + \mathbf{L}_{sek} = 0$$

$$\frac{\partial \zeta}{\partial t} + \nabla \cdot (D\mathbf{v}) = 0, \tag{3}$$

where ζ denotes the sea surface elevation with respect to the moving sea bottom, $\mathbf{v} = (u, v)$ the horizontal current velocity vector. The undisturbed ocean depth is D, the vector of Coriolis acceleration $\mathbf{f} = 2\omega \sin\phi \mathbf{z}$, the coefficient of linear bottom friction r' and the gravitational acceleration g. $\mathbf{F}$ denotes the vector defining the second-order eddy viscosity term $(F_\lambda, F_\phi) = (-A_h \Delta u, -A_h \Delta v)$ and (λ, ϕ) a set of geographic longitude and latitude values. The vector of the secondary force of the LSA (1) is

$$\mathbf{L}_{sek} = g\nabla \int G(\lambda, \phi, \lambda', \phi')\zeta(\lambda', \phi')d\lambda' d\phi' \cos\phi'.$$

The differential equation (3) is written as an eigenvalue problem

$$-i\mathscr{L}\bar{\mathbf{w}} = \sigma\bar{\mathbf{w}} \tag{4}$$

$$\mathbf{v}_H \cdot \mathbf{n}|_r = 0 \tag{5}$$

here, the periodic function $\mathbf{w}(\lambda, \phi) = \begin{pmatrix} \mathbf{v}_H \\ \zeta \end{pmatrix} = \begin{pmatrix} \bar{\mathbf{v}}_H \\ \bar{\zeta} \end{pmatrix} \cdot \exp(-i\sigma t)$, with the complex valued frequency $i\sigma = \sigma_1 + i\sigma_2$ is introduced. The operator $\mathscr{L}$, derived from the system of equations (3), is

$$\mathscr{L} = \begin{pmatrix} \mathbf{f} \times + \frac{r'}{D} - A_h \triangle_H & g\nabla_H - g\nabla_H \mathscr{I} \\ \nabla_H \cdot D & 0 \end{pmatrix}, \tag{6}$$

where $\mathscr{I}$ is defined through $\mathscr{I}\zeta = \int G(\lambda, \phi, \lambda', \phi')\zeta(\lambda', \phi')d\lambda' d\phi' \cos\phi'$.

Properly replacing the derivatives by finite differences and the integral by a finite expression [27] makes (4) turn into a system of algebraic equations

$$(A - \lambda I)\mathbf{x} = 0 \tag{7}$$

where $\lambda = i\sigma = \sigma_1 + i\sigma_2$ represents the eigenvalue of the Matrix A with the corresponding eigenvector $\bar{\mathbf{x}} = \mathbf{x}e^{-i\sigma t}$ depending on space as well as on time, i.e. $\bar{\mathbf{x}} = \bar{\mathbf{x}}(t, \lambda, \phi)$.

The system of equations (7) has, in the present case of a spatial resolution of one degree, approximately 120000 unknowns. Since LSA is taken fully into account the entries of the matrix A are generally nonzero. However, since the Green's function depends only on the angular distance a, symmetries in the arrangement of the entries can be utilized to reduce the working memory of the model to less than 1 GB (compare Eq. 2). Taking advantage of this memory reduction, three single free oscillations were computed with a special modification of the Wielandt method [28] and four with the standard Wielandt method [15]. In the first case the model was time optimized with respect to the method itself, whereas in the latter one it was distributed with OpenMP on 8 CPUs and optimized for the HLRE[1].

Both approaches make use of the Wielandt Method (or inverse iteration), originally developed by [24], and described in [6] or [8]. Starting from a first guess eigenvalue σ_0, the method yields the free oscillation with the eigenvalue λ closest to σ_0. The advantage of this method is that single free oscillations are determined with comparable low computational costs due to the possibility of the above mentioned memory reduction. The main disadvantage is the time consuming procedure when allowing for the full LSA-effect and that not all free oscillations are captured by this method.

3.2 The Implicitly Restarted Arnoldi Method

In the present study the Implicitly Restarted Arnoldi Method is used for solving the eigenvalue problem (4). It is provided by the software package ARPACK [11]. The original Arnoldi Method [1] is an orthogonal projection method, belonging to the class of Krylov subspace methods. In case of a symmetric matrix A, it reduces to the Lanczos Method [10]. Below, only a short summary of the Arnoldi method is given. A more comprehensive treatment of Krylov subspaces, Arnoldi factorization, and Arnoldi method can be found in [21].

The k-th Krylov subspace associated with the matrix A and the vector $\mathbf{v}$ is defined through

$$\mathscr{K}_k(A,\mathbf{v}) = \mathrm{span}\{\mathbf{v}, A\mathbf{v}, A^2\mathbf{v}, ..., A^{k-1}\mathbf{v}\} . \tag{8}$$

Obviously, it is defined through the sequence of vectors produced by the power method (e.g. [6]). This method utilizes the fact that with increasing k, the vector $A^k\mathbf{v}$ converges to the eigenvector with the largest eigenvalue. Like all Krylov subspace methods, the Arnoldi method takes advantage of the structure of the vectors produced by the power method, and information is extracted to enhance convergence to additional eigenvectors. For this purpose, the Arnoldi method determines an orthonormal basis $\mathrm{span}\{\mathbf{u}_1, \mathbf{u}_2, ...\mathbf{u}_k\}$ for $\mathscr{K}_k(A,\mathbf{v})$. This basis is defined through the relation

$$AU_k = U_k H_k + \mathbf{f}_k \mathbf{e}_k^T \tag{9}$$

where $A \in \mathbf{C}^{n\times n}$, the matrix $U_k = (\mathbf{u}_1, \mathbf{u}_2, ...\mathbf{u}_k) \in \mathbf{C}^{n\times k}$ (has orthogonal columns), $U_k^H \mathbf{f}_k = \mathbf{0}$, $\mathbf{e}_k \in \mathbf{C}^k$ and $H_k \in \mathbf{C}^{k\times k}$ is upper Hessenberg with non-negative subdiagonal elements. The algorithm of this so-called a k-step Arnoldi factorization is shown in Fig. 3.

[1] HLRE - High Performance Computing Center for Earth System Research, Hamburg

Input $(A, \mathbf{v})$

Put $\mathbf{u}_1 \mathbf{v}/\|\mathbf{v}\|$; $\quad \mathbf{w} = A\mathbf{u}_1$; $\quad \alpha_1 = \mathbf{u}_1^H \mathbf{w}$;

Put $\mathbf{f}_1 \leftarrow \mathbf{w} - \mathbf{u}_1 \alpha_1$; $\quad U_1 \leftarrow (\mathbf{u}_1)$; $\quad H_1 \leftarrow (\alpha_1)$

For $j = 1, 2, 3, \ldots k-1$

(1) $\beta_j = \|\mathbf{f}_j\|$; $\quad \mathbf{u}_{j+1} \leftarrow \mathbf{f}_j/\beta_j$;

(2) $U_{j+1} \leftarrow (U_j, \mathbf{u}_{j+1})$; $\quad \widehat{H}_j \leftarrow \begin{pmatrix} H_j \\ \beta_j \mathbf{e}_j^T \end{pmatrix}$;

(3) $\mathbf{w} \leftarrow A\mathbf{u}_{j+1}$;

(4) $\mathbf{h} \leftarrow U_{j+1}^H \mathbf{w}; \mathbf{f}_{j+1} \leftarrow \mathbf{w} - U_{j+1}\mathbf{h}$;

(5) $H_{j+1} \leftarrow (\widehat{H}_j, \mathbf{h})$;

End For

Fig. 3 Algorithm: The k-step Arnoldi Factorization

Alternatively, the factorization (9) can be written as

$$AU_k = (U_k, \mathbf{u}_{k+1}) \begin{pmatrix} H_k \\ \beta_k \mathbf{e}_k^T \end{pmatrix}, \tag{10}$$

where $\beta_k = \|\mathbf{f}_k\|$ and $\mathbf{u}_{k+1} = \frac{1}{\beta_k}\mathbf{f}_k$. If $H_k \mathbf{s} = \mathbf{s}\theta$, then the vector $\mathbf{x} = U_k \mathbf{s}$ satisfies

$$\|A\mathbf{x} - \mathbf{x}\theta\| = \|(AU_k - U_k H_k)\mathbf{s}\| = |\beta_k \mathbf{e}_k^T \mathbf{s}|. \tag{11}$$

The so called Ritz-pair $(\mathbf{x}, \theta)$ is an approximate eigenpair of A, with the Ritz-estimate as the residual $r(\mathbf{x}) = |\beta_k \mathbf{e}_k^T \mathbf{s}|$ (assuming $\|\mathbf{x}\| = 1$). Large amounts of memory are used to store all the basis vectors $\mathbf{u}_k$, if the number of iteration steps k is getting large before the eigenvalues and eigenvectors of interest are well approximated through the Ritz-pairs. Additionally, the computational cost of solving the Hessenberg eigenvalue sub-problem rises with $\mathcal{O}(k^3)$. To overcome these difficulties, methods have been developed to implicitly restart the method [23]. This efficient way to reduce the storage and computational requirements makes the Arnoldi Method suitable for large scale problems. Further, implicit restarting provides a means to determine a subset of the eigensystem. Hence, the ARPACK interface allows the user to specify the number l of eigenvalues sought.

When the Matrix A is considered in the Arnoldi method its l largest eigenvalues are determined. But in the case of the present study the interest lies in specific eigenvalues, e.g. those in the diurnal and semidiurnal spectrum. Therefore the shifted and inverted problem $(A - \sigma_0 I)^{-1}$ is considered. Thus the convergence of eigenvalues near the selected point σ_0 is enhanced. This approach is closely related to the inverse iteration techniques (e.g. [6]). Considering this spectral transformation in detail yields

$$A\mathbf{x} = \mathbf{x}\lambda \Longleftrightarrow (A - \sigma_0 I)\mathbf{x} = \mathbf{x}(\lambda - \sigma_0). \tag{12}$$

and

$$(A - \sigma_0 I)^{-1}\mathbf{x} = \mathbf{x}\nu, \; where \; \nu = \frac{1}{\lambda - \sigma_0} . \tag{13}$$

Hence, the eigenvalues λ that are close to σ_0 will be transformed into eigenvalues $\nu = \frac{1}{\lambda - \sigma_0}$, which are at the extremes of the transformed spectrum. The corresponding eigenvectors remain unchanged.

In case of the shifted and inverted approach of the Arnoldi Method, linear systems of the form $(A - \sigma_0 I)\mathbf{x} = b$ have to be solved. The algorithms of ARPACK are provided with a so called *reverse communication interface*. This interface allows the user to transfer the solution $\mathbf{x}$ into the algorithm, and in this way the solver can be chosen independently from ARPACK. In the present study, the LU-solver provided by ScaLAPACK [2] is used (see Sect. 3.3). The LU-solver puts itself forward since the time consuming LU-factorization of $(A - \sigma_0 I)$ need to be performed only once.

3.3 The Parallelization with MPI

To enable the use of routines of mathematical libraries for computing linear systems, it is necessary to store the complex matrix $(A - \sigma_0 I)$ in a general form. Thus the advantages of the symmetries of the matrix are getting lost. Since more than 500 GB of memory are required, it is necessary to parallelize the ocean model and distribute the matrix on different nodes. The parallelization is done with MPI[2], perfect for large problems needing access to large amounts of memory on distributed memory architectures [22].

The linear systems are solved with a parallelized version of a LU-solver of the ScaLAPACK software package [2]. Since the Matrix $(A - \sigma_0 I)$ is kept preserved during the whole iteration process of the Arnoldi algorithm, the LU-factorization, the most time consuming part, is only performed once. The choice of MPI and the ScaLAPACK LU-solver gives the user a high degree of freedom, in adapting the ocean model to the features of the computer architecture. The number of CPUs and nodes can freely be chosen, and is only restricted through the memory used to store the matrix.

3.4 The Performance of the Model

The model-runs have been performed on two distinct supercomputers, the HLRE[3] and the HLRS[4], equipped with NEC SX-6 nodes and NEC SX-8 nodes, respectively.

---

[2] Message-Passing Interface

[3] HLRE - High Performance Computing Center for Earth System Research, Hamburg

[4] HLRS - High Performance Computing Center, Stuttgart

Table 1 Data of the performance of the fastest model-run on 12 NEC SX-6 nodes of the HLRE: the first two rows show values from one single CPU; the last row are mean global values of all 96 CPUs.

	Frequency	Time in $[s]$	Performance in $[MFlops]$
LU-factorization (of CPU no. 1)	1	6872.6	6766.8
LU-solver (of CPU no. 1)	500	444.7	1373.1
Mean global values of 96 CPUs		8181.2	$608.7 \cdot 10^3$

Table 2 Data of the performance of model-runs on 4, 8, 16, 32 and 64 NEC SX-8 nodes of the HLRS.

Number of nodes	4	8	16	32	64
Number of CPUs	32	64	128	256	512
Real Time $[s]$	11421	6850	4463	3251	2766
Performance $[GFlops]$	416	737	1269	2108	3394

The number of free oscillations sought is set to $l = 150$ for each model-run. So 75 free oscillations and the corresponding complex conjugated ones are computed. At the HLRE, it is possible to run the program on 12 and 16 nodes. Each SX-6 is equipped with 8 CPUs. The overall performance is up to 609 and 632 GFlops, respectively, being one of the fastest single-application running on the HLRE. The computation time is between 2 and 3 hours, mostly depending on the actual state of the supercomputer and on the condition of the matrix, which changes through changing σ_0 value. Although the LU-factorization is highly optimized (see Table 1), it alone needs more than two third of the total time used by the model (in some cases up to 80%), the LU-solver uses 5-8%. The total memory of the model amounts to 630 GBytes.

Furthermore, model-runs have been performed on the HLRS supercomputer. The model was distributed on up to 512 CPUs (64 nodes). On this computer architecture, the good performance of the model is kept preserved (Table 2, Fig. 4), using only 45 minutes to determine 150 normal modes, with a mean performance of 3.4 TFlops.

4 Results

A total of 169 free oscillations is found in the period range of 7.7 h to 165 h [16]. The free oscillations of the World Ocean on the rotating Earth consist of gravitational and

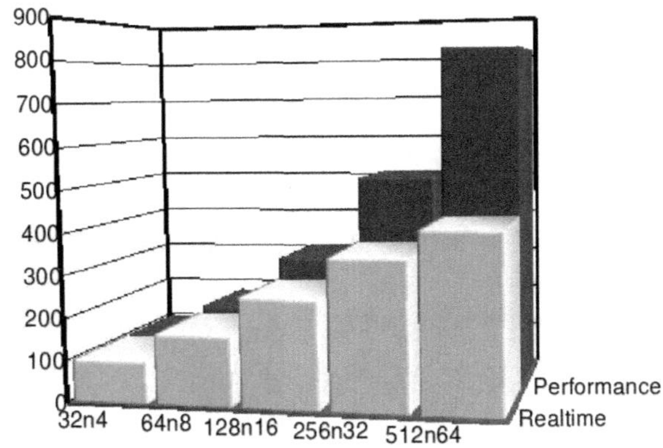

Fig. 4 The speed up of performance and real time of the ocean in %, on 32, 64, 256 and 512 CPUs.

vorticity modes, primarily governed by the gravity of the Earth and by the latitude dependent Coriolis force, respectively.

4.1 Gravitational Modes

Two of the 72 gravitational modes, are exemplarily shown in Fig. 5. The 11.98-mode has its main energy located in the Pacific. There, it appears mainly as a Kelvin Wave around New Zealand, which is a prominent feature of the semidiurnal tides and indeed this mode plays an important role in the synthesis of the semidiurnal tides. The 11.98-mode resembles the mode 38 (10.8 h, Fig. 20) of [18] who named it the "free New Zealand Kelvin Wave". In the 32.64-mode almost all energy is located in the Southern Ocean. There a wave travels in westward direction and trapped by the Antarctic and, it is accompanied by a strong westward energy flux. It is called the Antarctic Kelvin Waves of orders of one, defined through the number of wavelengths within one cycle. The 32.64-mode is an important mode for the diurnal tides although the period is not close to the forcing periods. This is due to the large coherence of the diurnal potential with the adjoint mode of the 32.64-mode [17].

4.2 Vorticity Modes

In total, 91 of the 97 determined vorticity modes are topographical modes. They are trapped by prominent topographic structures over which they focus the main part of their total energy. They are characterized by an energy-flux gyre and a strong modification of their relative vorticity with changing water depth (more precisely: normal to the contours of $\frac{f}{h}$). The behavior of the transport ellipses is characterized by an inversion of the sense of rotation at a certain depth contour line. The rotation is anti-cyclonic over the topographic structure and cyclonic in deeper water,

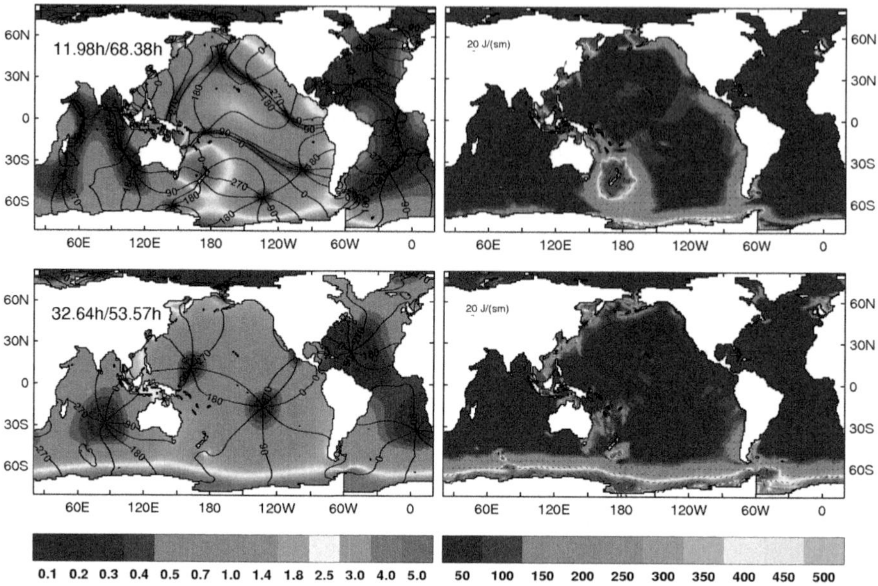

Fig. 5 Gravitational modes: **Left** normalized amplitudes (color contoured) of sea-surface elevation and lines of equal phases. **Right** energy flux vectors for every fifth grid point, zonally and meridional. Squaring the magnitudes given yield these quantities in J/(sm). The energy flux vector at the top left has a magnitude of 20^2 J/(sm). The magnitude of the energy flux is additionally color contoured. **Top** the 11.98-mode with a decay time of 68.38 h. **Bottom** the 32.64-mode with a decay time of 53.57 h.

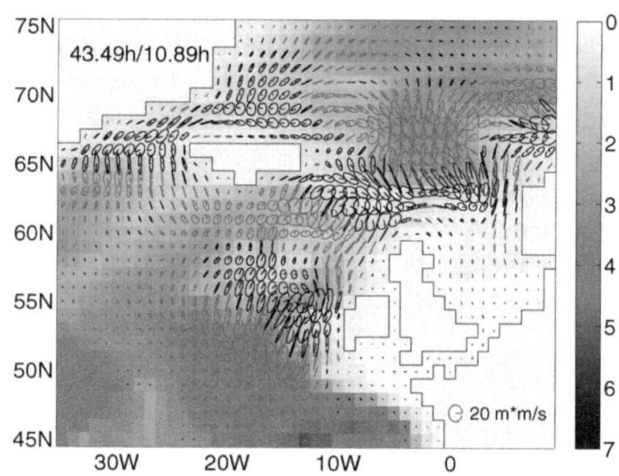

Fig. 6 Topographical mode: Volume transport ellipses plotted for the topographical vorticity mode in the North Atlantic with a period of 43.49 hours and a decay time of 10.89 hours. Red ellipses indicate cyclonic rotation, black ones anticyclonic rotation. The bathymetry is color-contoured and labeled in [km].

consistent with previous observations and numerical models (e.g. [9]). In Fig. 6 an topographical mode centered at the Faeroe-Island Ridge is shown. It resembles the mode that was measured and analyzed by [14].

Altogether, six planetary normal modes have been computed with periods of 96.94, 111.22, 113.44, 119.65, 136.87, and 141.66 h. Five of them have a global distribution and one (113.44 h) is limited to the Pacific Ocean. The planetary mode

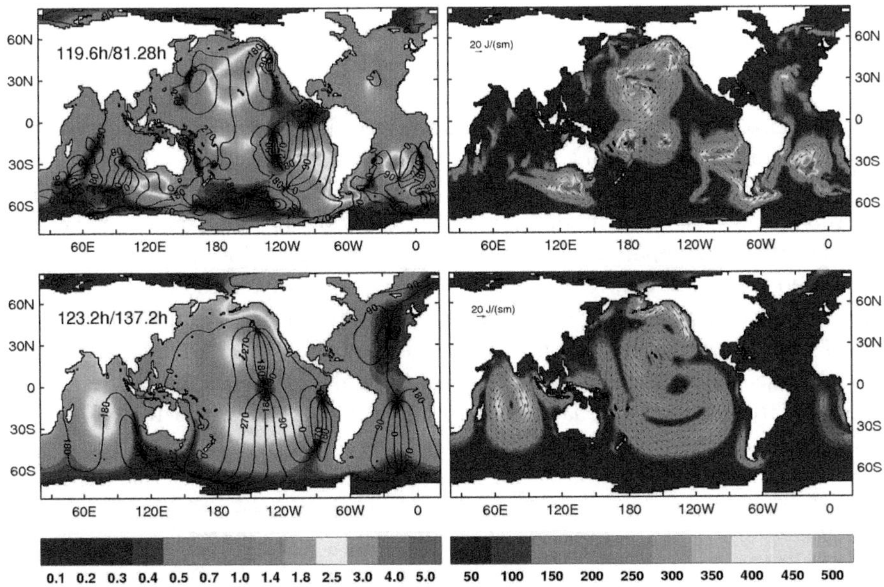

Fig. 7 Planetary modes: **Left** normalized amplitudes (color contoured) of sea-surface elevation and lines of equal phases. **Right** energy flux vectors for every fifth grid point, zonally and meridionally. Squaring the magnitudes given yield these quantities in J/(sm). The energy flux vector at the top left has a magnitude of 20^2 J/(sm). **Top** the 119.6-mode with a decay time of 81.28 h. **Bottom** the corresponding mode calculated with constant depth ($D = 4000$ m).

with a period of 119.65 h (Fig. 7) has a global energy distribution with 13% of E_t in the Indian and Southern Ocean through topographical trapping at the Southeast Indian Ridge. Of the total energy, 20% is located in the Atlantic, mainly in a resonance in the middle of the South Atlantic, and 67% in the Pacific. The structure in the Pacific is similar to that of [13] (see his Fig. 4d) with a period of 120 h, but it appears in the 119.65-mode meridionally compressed and bounded in the east by the East Pacific Rise. An additional resonance east of the East Pacific Rise arises from a planetary wave with half wavelength. The energy flux (Fig. 7) shows that the ridge reflects the energy and, thus, is responsible for this sub-basin (bounded in the east by South America) resonance. Part of this energy is transported through the Drake Passage in the Atlantic. Further, in Fig. 7 the corresponding mode determined with flattened ocean of 4000 m depth is shown. It reflects that the planetary modes are strongly influenced by the bottom topography, either through trapping or by functioning as a barrier.

5 Conclusion

The objective of this study was to compute a large spectrum of free oscillations of a linear, barotropic, one degree global ocean model with explicit consideration of dissipative terms and the full LSA-effect. This task could be coped with by utilizing the Implicitly Restarted Arnoldi Method and implementing it in a specific way on a supercomputer. The ocean model is constructed to run in parallel on a large number of CPUs of supercomputers with a shared memory distribution. The parallelization is required to allow the usage of large amounts of memory, needed for this kind of eigenvalue problem. With the developed ocean model, the highly efficient computation of large spectra of free oceanic oscillations with the full LSA-effect, is enabled. Thus, the LSA-effect, which so far appears to be the greatest unknown left in the barotropic free oscillation behavior of the open ocean, can now be better understood and quantitatively estimated.

References

1. Arnoldi, W.E.: The principle of minimized iterations in the solution of the matrix eigenvalue problem. Quaterly of Applied Mathematics **9**, 17–29, (1951).
2. Blackford, L.S., Choi, J., Cleary, A., D'Azevedo, E., Demmel, J., Dhillon, I., Dongarra, J., Hammarling, S., Henry, G., Petitet, A., Stanley, K., Walker, D., Whaley, R.C.: Scalapack users' guide. SIAM (1997)
3. Farrell, W.E.: Deformation of the earth by surface loads. Rev. Geophys. Space Phys. **10**, 761–797 (1972)
4. Francis, O., Mazzega, P.: Global charts of of ocean tide loading effects. J. Geophys. Res. **95**, 411–424 (1990)
5. Gaviño, J.H.R.: Über die Bestimmung von reibungslosen barotropen Eigenschwingungen des Weltozeans mittels der Lanczos-Methode. PhD thesis, University of Hamburg (1981)
6. Golub G., van Loan, C.: Matrix Computations, 2 edn. Johns Hopkins University Press (1989)
7. Gotlib, V.Y., Kagan, B.A.: On the resonance excitation of semidiurnal tides in the world ocean. Fizika atmosfery i okeana **17**, 502–512 (1981)
8. Ipsen, I.C.F.: A history of inverse iteration. In: B. Huppert, H. Schneider (eds.) Helmut Wielandt, Mathematische Werke, Mathematical Works, II: Matrix Theory and Analysis, pp. 464–472. Walter de Gruyter, New York (1996)
9. Kowalik, Z.: Modeling of topographically amplified diurnal tides in the nordic seas. J. Phys. Oceanogr. **24**, 1717–1731 (1994)
10. Lanczos, C.: An iteration method for the solution of the eigenvalue problem of linear differential and integral operators. Journal of Research of the National Bureau of Standards **45**, 255–282 (1950). Research Paper 2133
11. Lehoucq, R.B., Sorensen, D.C., Yang, C.: ARPACK user's guide: Solution of large scale eigenvalue problems by implicitly restarted arnoldi methods. Technical report, Department of Computational and Applied Mathematics, Rice University (1996)
12. Longuet-Higgins, M.S.: The eigenfunctions of laplace's tidal equations over a sphere. Phil. Trans. Roy. Soc. London **A262**, 511–581 (1968)
13. Miller, A.J.: On the barotropic planetary oscillations of the pacific. J. Mar. Res. **47**, 569–594 (1989)
14. Miller, A.J., Lermusiaux, P.F.J., Poulain, P.: A topographic-rossby mode resonance over the iceland-faeroe ridge. J. Phys. Oceanogr. **26**, 2735–2747 (1996)

15. Müller, M.: Barotrope Eigenschwingungen im Globalen Ozean unter Berücksichtigung des vollen Selbstanziehungs- und Auflasteffektes. Master's thesis, University of Hamburg (2003)
16. Müller, M.: The free oscillations of the world ocean in the period range 8 to 165 hours including the full loading effect. Geophys. Res. Lett. **34** (L05606) (2007)
17. Müller, M.: Synthesis of forced oscillations, Part 1: Tidal dynamics and the influence of LSA Ocean Modelling. **20**, 207–222 (2008)
18. Platzman, G.W., Curtis, G.A., Hansen, K.S., Slater, R.D.: Normal modes of the world ocean, Part 2: Description of modes in the range 8 to 80 hours. J. Phys. Oceanogr. **11**, 579–603 (1981)
19. Rao, D.B.: Free gravitational oscillations in rotating rectangular basins. J. Fluid Mech. **25**, 523–555 (1966)
20. Ray, R.D.: Ocean self-attraction and loading in numerical tidal models. Mar. Geod. **21**, 181–192 (1998)
21. Saad, Y.: Numerical methods for large eigenvalue problems: theory and algorithms. John Wiley, New York (1992)
22. Snir, M., Otto, S., Huss-Lederman, S., Walker, D., Dongarra, J.: MPI — The complete reference vol. 1: The MPI core. MIT Press (1998)
23. Sorensen, D.C.: Implicit application of polynomial filters in a k-step Arnoldi method. SIAM Journal on Matrix Analysis and Applications **13**, 357–385 (1992)
24. Wielandt, H.: Das Iterationsverfahren bei nicht selbstadjungierten linearen Eigenwertaufgaben. Math. Z. **50**, 93–143 (1944)
25. Zahel, W.: Mathematical modelling of global interaction between ocean tides and earth tides. Phys. Earth Planet. Interior **21**, 202–217 (1980)
26. Zahel, W.: Astronomical tides. In: J. Sündermann (ed.) Numerical Data and Functional Relationships in Science and Technology, vol. 3c of Landolt–Börnstein, chap. 6.4, pp. 83–134. Springer-Verlag (1986)
27. Zahel, W.: Modeling ocean tides with and without assimilating data. J. Geophys. Res. **12**, 20379–20391 (1991)
28. Zahel, W., Müller, M.: The computation of the free barotropic oscillations of a global ocean model including friction and loading effects. Ocean Dynamics **55**, 137–161 (2005)

Direct Numerical Simulation of Controlled Shear Flows

Markus J. Kloker, Tillmann A. Friederich, Jens Linn

Abstract Two examples of controlled boundary-layer flows are discussed to illustrate the potential of advanced shear-flow control for drag or heat-load reduction. Delay of the drag-increasing laminar-turbulent transition, a procedure generally termed laminar flow control (LFC), is demonstrated in a three-dimensional boundary-layer as present on a swept airliner wing. Using a code for incompressible flow localized suction at the wall is applied at steady crossflow vortices exactly below the region where a strong localized high-frequency instability typically triggers turbulence, and significant transition delay is demonstrated at relatively low suction rates. Using a compressible code, effusion cooling through holes in the laminar boundary layer of a vehicle flying at Mach= 6.8 in the atmosphere is simulated and investigated with respect to cooling efficiency as well as possible destabilization of the laminar flow, the latter possibly destroying the benefit of cooling. In both cases true unsteady simulations are performed to be able to correctly detect and capture flow instabilities, rendering the simulations costly. The performance of the used codes is discussed as well as the problems that will be attacked in the near future.

1 Laminar-Flow-Control Case

1.1 Introduction

The substantially grown costs for fuel nowadays amount up to 40% of the direct operating costs of a long-range flight. So even before environmental protection laws enforce a decrease in exhaust gases there is vital interest in lowering fuel consumption by lowering the aerodynamic drag of airliners. As the viscous drag share during

Markus J. Kloker · Tillmann A. Friederich · Jens Linn
Institut für Aerodynamik und Gasdynamik, Universität Stuttgart, Pfaffenwaldring 21, 70569 Stuttgart, Germany, e-mail: {kloker, friederich, linn}@iag.uni-stuttgart.de

cruise is about 50%, its reduction offers the largest potential for fuel savings. This can be achieved with laminar flow control (LFC) by boundary-layer suction on the wings, tailplanes, and nacelles with a fuel saving potential of about 16% if laminar flow could be kept up to 40% chord of the respective surface. The management of turbulent flow, e.g., on the fuselage of the aircraft, by a kind of shark-skin surface structure has a much lower total saving potential, of about 1–3% with treated fuselage.

Boundary-layer suction has been known for long to significantly increase laminar stability and thus to push laminar-turbulent transition downstream. In case of swept aerodynamic surfaces with three-dimensional boundary layers suction aims primarily at reducing the inherent, nocent crossflow within the boundary layer by sucking fluid from the outer region of the boundary layer with higher momentum to the wall. The crossflow causes a primary instability of the boundary layer typically leading to co-rotating longitudinal vortices, called crossflow vortices (CFVs), that typically trigger a high-frequency secondary instability invoking the substantially drag-increasing turbulence. Discrete suction through a perforated wall to diminish primary instability can excite nocent crossflow vortices, but when the perforations are smartly arranged, they can excite useful crossflow vortices that lie closer together and suppress the growth of other disturbances. This technique is called distributed flow deformation (DFD) and has been proposed by Messing & Kloker (see, e.g., [11]).

With deliberately excited vortices, whose non linearly large amplitudes vary downstream with varying baseflow, it may on the other hand be necessary to directly control their stability and thus to secure or enlarge their range of usefulness. The aim is then to prevent or weaken localized (secondary) instability of vortex-deformed flow. To this end, very localized suction is applied directly below the possibly dangerous, localized shear layers invoked by a (crossflow) vortex that transports low-momentum fluid from the wall to the boundary-layer edge inducing streamwise (and wall-normal) vorticity. This is investigated here in the canonical three-dimensional boundary-layer flow of the DLR-Göttingen-Prinzipexperiment (see Bippes [2]) as also used by Bonfigli & Kloker [3], where the baseflow is thoroughly described. The idea to use this specialized suction was fostered by the finding that even a small velocity component normal to a local shear layer can substantially reduce the shear-layer instability (Bonfigli & Kloker [3], Friederich [4]).

The high-frequency secondary instabilities of crossflow vortices (CFVs) has been fully elucidated in the last few years experimentally (White & Saric [14]), theoretically (secondary linear stability theory (SLST), Koch & al. [7]), and by means of spatial DNS (Wassermann & Kloker [13]), see also the overview by Saric, Reed, and White [12]. Quantitative comparisons showed significant deviations between amplification rates from DNS and SLST. The diversity could be traced back to a sensitivity of secondary growth rates with respect to very small changes in the primary state [3], caused by different simplifying representations of the same primary flow field.

Thus the secondary instability mechanisms are assumed to be highly susceptible to moderate suction or blowing at the wall, providing a good chance to delay transition to turbulence. By means of spatial DNS, we show that the amplitude growth rates of artificially excited unsteady high-frequency secondary instabilities can indeed be reduced drastically by exploiting the shear-layer sensitivity with respect to a normal velocity component.

1.2 Numerical Model

The code N3D solves the full 3D unsteady incompressible Navier-Stokes equations using a disturbance formulation, where each flow quantity q is divided into its steady baseflow part q_B and its unsteady component q' to ease formulation of the boundary-conditions. We focus on secondary instabilities, thus our primary state consists of the baseflow plus steady CFVs. For the downstream and wall-normal directions x and y sixth-order compact splitted finite differences are implemented whereas for the spanwise direction z, we use a Fourier spectral representation with $K+1$ modes. A sketch of the integration domain is shown in Fig. 1, and for a detailed description of the numerical method see [13] and [3].

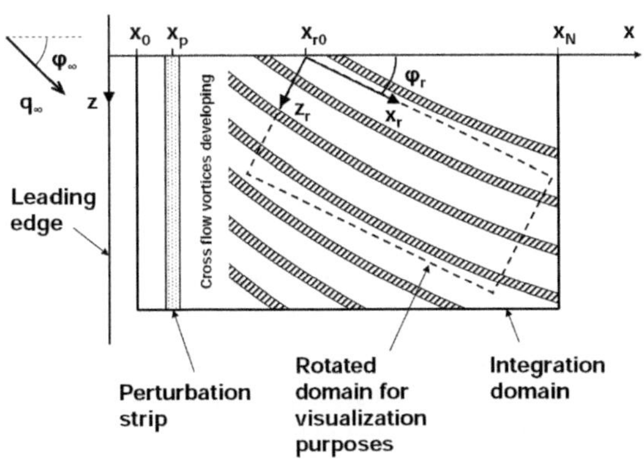

Fig. 1 Flat plate (top view) with integration domain and rotated reference system (x_r, z_r) used for visualization purposes

We use the baseflow corresponding to the DLR-Göttingen experiment [2, 3], where a flat-plate flow with negative streamwise pressure gradient was generated by a displacement body above. For all simulations the free-stream velocity $q^*_\infty = 19\,\text{m/s}$ and a sweep angle of $\Phi_\infty = 42.5°$ were chosen. Thus, the component $U^*_\infty = 14\,\text{m/s}$ and a reference length of $L^* = 0.1\,\text{m}$ are used for non-dimensionalization. The Reynolds number is $Re = 92000$ based on U^*_∞ and L^*. The extensions of the integration domain in streamwise and wall-normal direction are $l_x = 0.219\,\text{m}$ (1674 points) and $l_y = 0.014\,\text{m}$ (225 points, step size decreases while approaching the wall). The fundamental wavelength in spanwise direction $\lambda_{z,0} = 0.012\,\text{m}$ is discretized with $K = 23$ harmonics and the fundamental reference frequency is $\omega_0 = 6$ ($f^* = 133\,\mathrm{Hz}$).

1.3 Secondary Instability and Control Setup

At $x = 3.0$ we excite unsteady low-amplitude disturbances with $\omega = h \times \omega_0$, $h = 1-50$ and $\gamma = \pm\gamma_0 = \pm 52.4$. Figure 3 shows the downstream modal amplitude development of the streamline-oriented velocity component $\tilde{u_s}' = \frac{u_s'}{u_{B,s,e}}$ of the secondary modes.

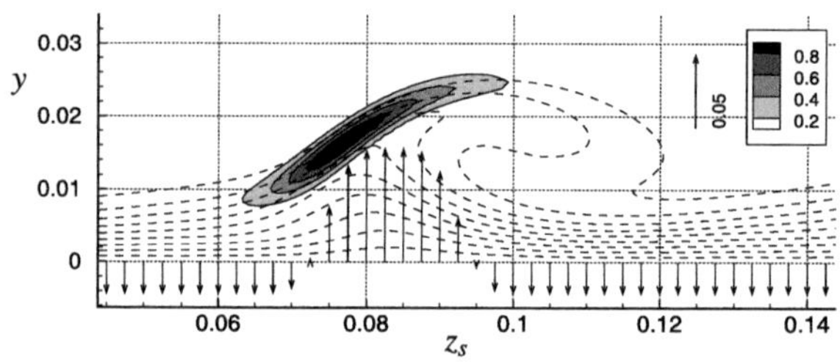

Fig. 2 Normalized amplitude distribution for the mode with angular frequency $\omega = 132$ (shaded), normalized u_s-velocity isocontours (dashed lines) and synthetic blowing pattern (arrows, x-extension $l_x \approx 0.05$) in a flow crosscut at $x = 3.32$

Still within the linear stage ($x < 3.8$), high-frequency modes achieve both the highest amplitudes and the highest growth rates ($x > 3.4$). Soon downstream at $x = 3.8$, explosive non-linear growth leads to first stages of transition ($x > 4.2$), indicated by the decreasing steady 3D part of the primary disturbance $\omega = 0-(0,0)$ and the strong deformation of vortical structures in a λ_2-visualization plot (cf. Fig. 7) which shows a rotated snapshot of the integration domain including four co-rotating CFVs. A more detailed investigation of mode $\omega = 132$ reveals that the corresponding amplitude distribution shows the typical S-shaped type-I mode as plotted in Fig. 2. In the scenario here this mode is known to be responsible for transition to turbulence by generating secondary finger-like vortical structures at the left, updraft side of the main vortex (when looking in downstream direction). Therefore, synthetic blowing (see Fig. 2, arrows), with zero net mass flow, and non-synthetic suction patterns (not shown) are imposed at the wall such that the wall-normal velocity peaks lie right underneath the amplitude maxima of the unstable modes. Their spanwise positions can be easily located by identifying the minimum of the wall-normal gradient of the primary-state streamwise velocity component u. Two major effects are expected to influence the flow field: on the one hand, in the case of blowing, a local strengthening of the primary vortex can cause a more unstable scenario with respect to secondary instabilities, and for a suction scenario, we expect a locally weakened primary vortex and thus less unstable flow conditions. On the other

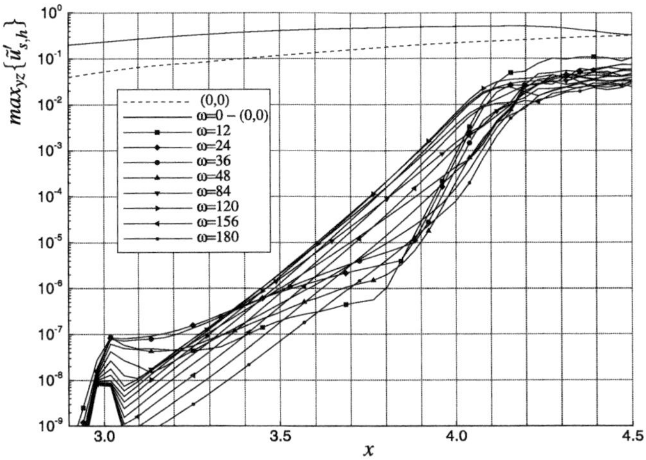

Fig. 3 Downstream development of t-modal amplitudes $\tilde{u}_s'$ (maximum over y and z) for the reference case without blowing or suction. Unsteady wave-packet disturbances are excited at $x = 3.0$

hand, due to the sensitivity of the secondary modes, the alteration of the wall-normal velocity component v should damp their spatial amplitude growth significantly and overcompensate the previous effect.

1.4 Control Results and Conclusions

The amplitude development of the u_s-velocity component for the synthetic blowing case is provided in Fig. 4. Starting at $x = 3.3$, the expected strengthening of the main vortex can be observed in the steady 3D component of the main vortex.

However, a damping effect on the growth rates of the unstable secondary modes ($x > 3.5$) results in damped amplitudes and delayed transition. The sensitivity of the secondary modes with respect to minor changes in the wall-normal velocity component is successfully exploited and the predicted mechanisms seem to provide stronger damping effects than the disadvantageous locally strengthening of the main vortex.

For the suction case (cf. Fig. 5) an almost stagnating amplitude growth can be observed up to $x = 4.1$ and transition is prevented inside the considered domain. The development of the type-I mode downstream of the suction strip reveals a stretching and weakening effect on the mode (Fig. 6) which leads to smaller amplitude growth and hence a longer laminar regime. Figure 8 reveals (throughout the integration domain almost) undisturbed CFVs, whereas for the reference case (Fig. 7), at the same time level, strong deformation of the vortical structures can be observed that will soon trigger transition to turbulence.

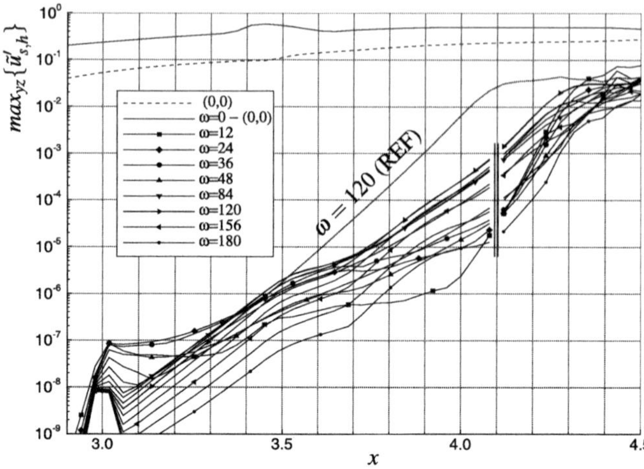

Fig. 4 Downstream development of modal amplitudes $\tilde{u}_s{}'$ (maximum over y and z) for synthetic blowing at $x = 3.37$ according to Fig. 2 ($v_{max,wall} = 7.8\%$). The curves left of the two vertical lines result from a technique that analyzes the second temporal derivative in order to avoid Fourier analysis problems with slightly unsteady flow fields, see [9].

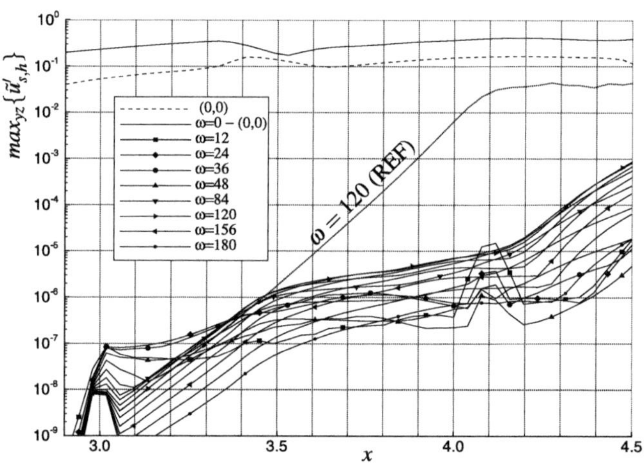

Fig. 5 Downstream development of modal amplitudes $\tilde{u}_s{}'$ (maximum over y and z) for real suction at $x = 3.37$ ($v_{max,wall} = 10\%$). Second temporal derivative technique has been applied, see [9]

Further investigations show that (in spanwise direction) misplaced blowing or suction holes can create strong shear layers that may even amplify the amplitude growth rates of unsteady secondary modes. A well-positioned hole pattern that is adapted to the local flow characteristics, especially the vortex situation, is therefore indispensable.

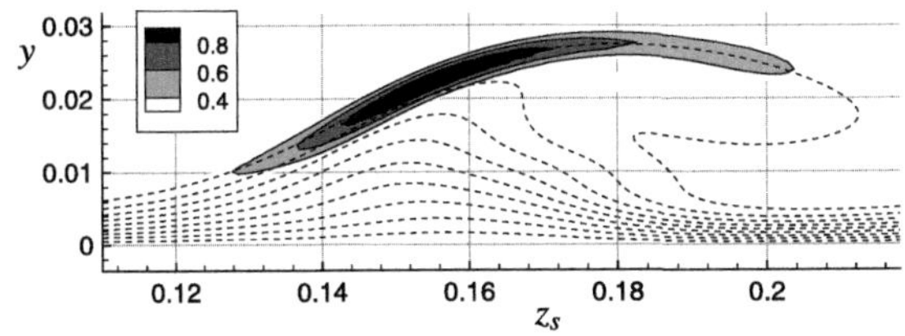

Fig. 6 Normalized amplitude distribution for mode with frequency $\omega = 132$ (shaded) and u_s-velocity isocontours (dashed lines) for the suction case in crosscut at $x = 3.92$

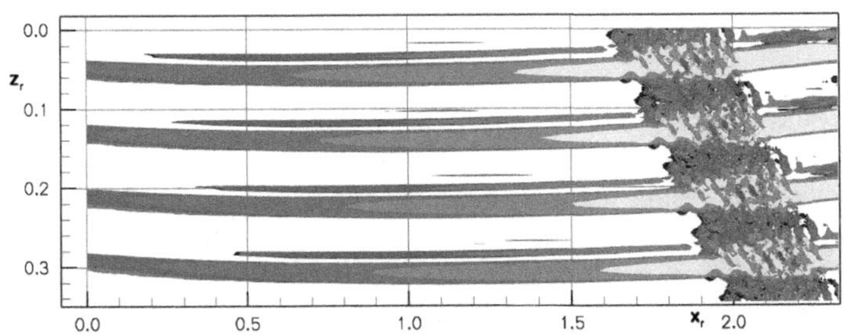

Fig. 7 Top view on plate: λ_2-visualization for the reference case, snapshot at $t/T_0 = 0.0$, $\lambda_2 = -10$. A rotated reference system has been used for visualization: $x_{r,0} = 2.79, z_{r,0} = 0.0, \Phi_r = 45.86°$. The domain covers approx. $x \in [2.6, 4.5]$. Four fundamental periods in z_r-direction are shown. The color indicates the wall-normal distance, where blue is $y = 0.00$, red $y = 0.04$, and green and yellow intermediate.

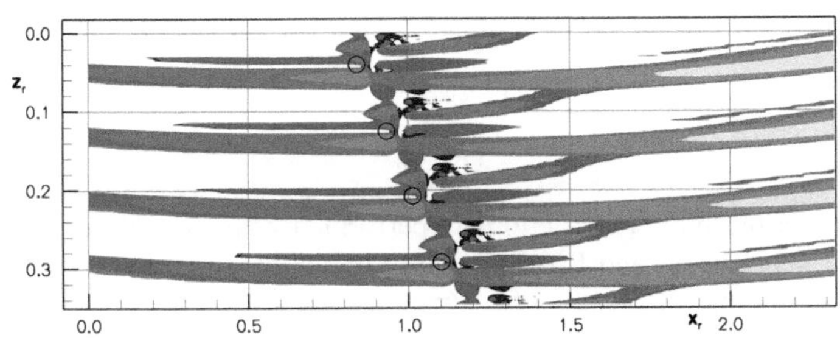

Fig. 8 Top view on plate: λ_2-visualization for the case with suction, snapshot at $t/T_0 = 0.0$, $\lambda_2 = -10$. Circles mark locations of the suction holes. Same reference system as in Fig. 7.

The results show that it is possible to exert a weakening influence on highly unstable secondary modes by accurately positioned moderate suction and/or blowing, thus delaying transition to turbulence. In a scenario with localized wall-normal jets it was shown that the damping effect on the growth rates of unsteady modes could surpass the unwanted local strengthening of the main vortex. Significant damping was observed for the suction scenario where not only the attenuation effect on the unstable mode but also the influence on the primary state (i.e. both reducing the crossflow component and weakening the main vortex) lead to the largest amplitude reduction of more than 99%. The needed suction rate, i.e. the suction velocity averaged over the surface area that contains the perforations, is by a factor of 3–10 lower than for "primary" suction. More details can be found in Friederich [4] and Friederich & Kloker [5].

2 Effusion Cooling Case at <Mach 6.8> Flight Conditions

2.1 Introduction

For hypersonic cruise or aerospace vehicles the heat load due to friction is critically high and depends on the boundary-layer state that can be laminar, transitional, or turbulent, with increasing load. "Simple" cooling by radiation of heat is not sufficient even for cruise vehicles, and additional cooling by ablation of material or transpiration or effusion of a cooling fluid or gas is necessary. Here, direct numerical simulations are carried out to investigate the effect of effusion cooling by blowing through spanwise slits and discrete holes onto a laminar, radiation-cooled flat-plate boundary layer developing at a flow Mach number of 6.8. The shear and temperature layers induced by the coolant flow may destabilize the laminar flow leading to early transition to turbulence, partially compromising the cooling effect. Thus knowledge of the detailed flow and temperature field with active cooling is essential for the reliable design of the thermal protection system.

2.2 Numerical Model

The numerical results in this section are performed with the DNS code NS3D (see Babucke, Linn, Kloker, and Rist [1]). This code is based on the complete 3D unsteady compressible Navier-Stokes equations and a calorically perfect gas. The equation set is solved in a rectangular integration domain on the flat plate, well below the shock wave induced by the leading edge. In streamwise (x-) and wall-normal (y-) direction, the discretization is realized by splitted compact finite differences of 6th order. In the spanwise (z-) direction, the flow is assumed to be periodic, thus a Fourier spectral representation is employed to compute the z-derivatives. In contrast to the incompressible code N3D, NS3D largely computes in physical space. After transformation to Fourier space and simple computation of the z-derivatives, the back transformation is done with de-aliasing using the 2/3-rule. For time integration the classical 4th-order Runge-Kutta method is employed as in the incompressible case. A detailed description of the discretization algorithm, and boundary conditions is given in Babucke & al. [1] and Linn & Kloker [8].

2.3 Blowing Through Slits

In this section we investigate an effusion-cooled Mach 6.8 boundary layer at flight conditions at an altitude of 33 km, corresponding to a flight point of the hypersonic "Sänger" lower-stage vehicle [6]. We prescribe a radiation-adiabatic wall and $T^\star_\infty = 231.5$ K, thus we would have $T^\star_{rec} = 2031$ K ($Pr = 0.706$) for an adiabatic wall without radiation. Recall that the radiation-adiabatic condition means that there is no effective heat flux into the vehicle at a temperature lower than $T^\star_{rec}$, because the heat flux from the gas into the vehicle surface is radiated ($q^\star_{rad} = \varepsilon\, \sigma\, T^{\star^4}_w$). With a surface emissivity coefficient $\varepsilon = 0.8$ we get $T^\star_w \approx 0.4\, T^\star_{rec} \approx 3.5\, T^\star_\infty$, corresponding to 930 K at $x = 1$, decreasing to 750 K at $x = 9$ without blowing, $p^\star_\infty = 0.00558$ bar, $L^\star = 85.367$ mm, and $Re^\star_\infty = 1.17 \times 10^6$ 1/m. In cases $\mathscr{A}$ and $\mathscr{B}$ cool air is blown through two successive slits into the boundary layer. The geometrical data of the slits are listed in Table 1. The blowing through slits is modeled by prescribing a mass flux $(\rho v)_c$ and a cooling-gas temperature T_c distribution at the wall [8]. The cooling gas temperature is $T^\star_{c,core} = 293$ K ($\approx 0.14\, T^\star_{rec}$) and the integrally injected mass flow is the same in both cases. All dimensional quantities have the superscript $\star$ and all other quantities are normalized with their values in the oncoming flow or with reference values. Especially, $(\rho v)_c = (\rho v)^\star_{cool}/(\rho u)^\star_\infty$.

Table 1 Parameters of the slit configurations for cases at $M_e = 6.8$

case	$(\rho u)_{c,max}$	$(\rho v)_{c,max}$	blowing angle ϕ_c	slit width d	streamwise spacing s_x
$\mathscr{A}$	—	0.0283	0°	0.0234 = 2 mm	0.0585
$\mathscr{B}$	0.02	0.02	45°	0.033 = 2.83 mm	0.0585

In case $\mathscr{B}$ inclined blowing by an inclined duct is applied, see Fig. 9. The slit width $d_{inclined}$ $(= d/\cos\phi)$ is larger and the wall-normal mass flux (per unit area) $(\rho v)_{c,max} = (\rho v)_{c,max,0^\circ} \cos\phi$ is lower than in case $\mathscr{A}$. Of course, the injected mass flow is identical, only the blowing area is enlarged and the wall-normal mass flux lowered. However we additionally have a wall-parallel flux at the wall, somehow like a locally moving, cold, permeable wall within the slits compared to the standard case.

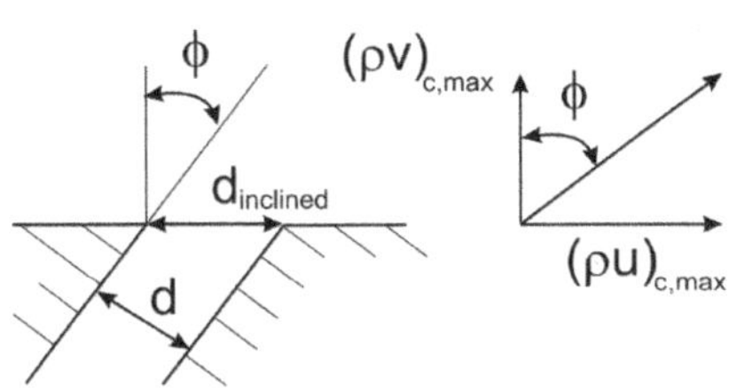

Fig. 9 Sketch of the inclined blowing

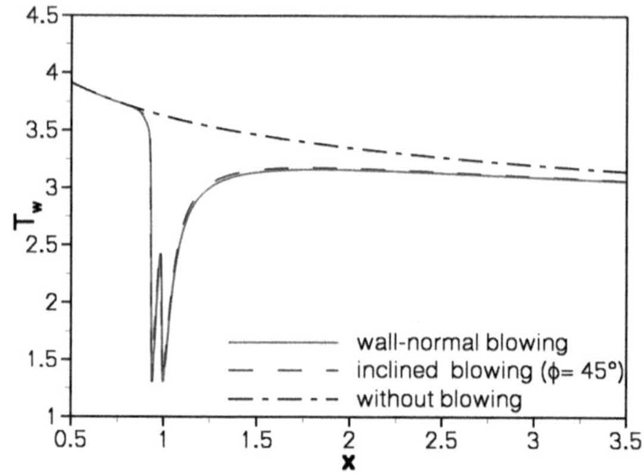

Fig. 10 Streamwise wall-temperature evolution for blowing through two slits (cases $\mathscr{A}$ and $\mathscr{B}$ of Table 1) for a radiation-adiabatic Mach-6.8 boundary layer

The wall-temperature distribution is shown in Fig. 10. The cases with wall-normal ($\mathscr{A}$) and inclined blowing ($\mathscr{B}$) look similar. The reason may be the low blowing rates. For higher blowing rates we expect the inclined blowing to yield a lower wall temperature. Note that the slit width is about 0.5δ (δ — boundary-layer thickness) for case $\mathscr{A}$, and that 10δ downstream of the second slit ($x = 1.5$) the wall is still cooled by about 70 K despite the small injected mass flow of about 3% of the total boundary-layer mass flow.

Figure 11 shows longitudinal cuts of the temperature field for both cases. At first sight the temperature fields seem similar, too. Only near the slits the temperature is slightly lower in case $\mathscr{A}$. The streamlines coming out of the slits show a lower angle than 45° in case $\mathscr{B}$ due to the stretched y-coordinate.

In Fig. 12 the downstream evolution of the cooling effectiveness η by blowing through the two slits is shown for wind tunnel conditions (WTC) and for case $\mathscr{A}$ at flight conditions (FC). WTC here means that $T^{\star}_{\infty,FC} = 89$ K with an adiabatic wall, simulating an experiment in the H2K tunnel of DLR in Cologne (see also [8]). The cooling effectiveness η is defined by

$$\text{WTC}: \quad \eta_{ad} = \frac{T^{\star}_{rec} - T^{\star}_{w,c}}{T^{\star}_{rec} - T^{\star}_{c,core}} \tag{1}$$

and

$$\text{FC}: \quad \eta_{rad} = \frac{T^{\star}_{w,rad} - T^{\star}_{w,c}}{T^{\star}_{w,rad} - T^{\star}_{c,core}}, \tag{2}$$

where $T^{\star}_{rec}$ is the recovery temperature, $T^{\star}_{w,rad}$ is the local wall temperature without blowing, and $T^{\star}_{w,c}$ is the local wall temperature with blowing. Due to the blowing of cold air into the boundary layer, the wall temperature decreases, which results in a partial loss of the radiation cooling ($q^{\star}_{rad} = \varepsilon\, \sigma\, T^{\star 4}_{w}$) at FC. Thus the cooling effectiveness in this case is lower than at wind tunnel conditions.

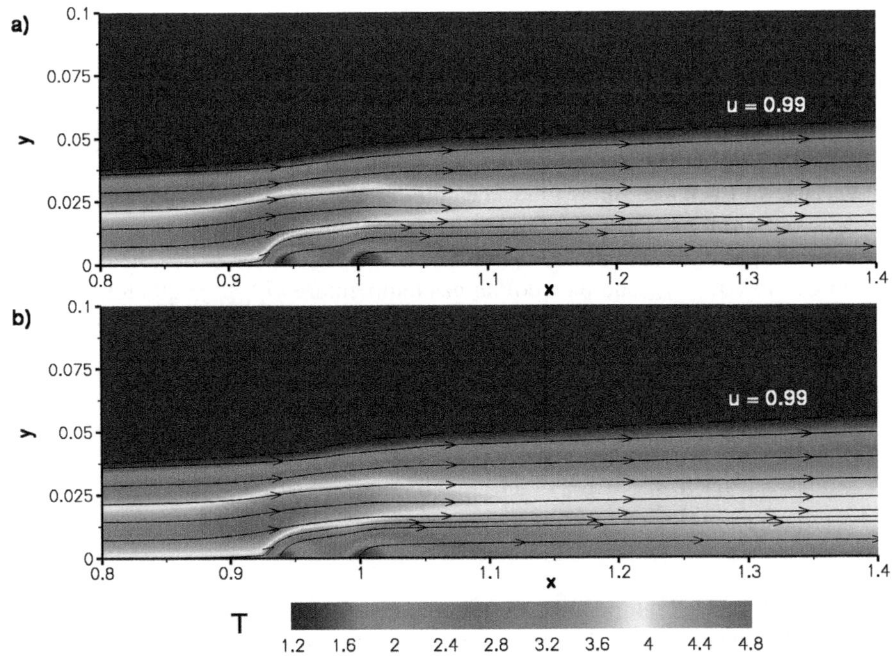

Fig. 11 Temperature fields and streamlines in a longitudinal cut for the radiation-adiabatic Mach-6.8 boundary layer with wall-normal blowing (case $\mathscr{A}$) and inclined blowing (case $\mathscr{B}$) — see Table 1). Dashed-dotted line: $u = 0.99$.

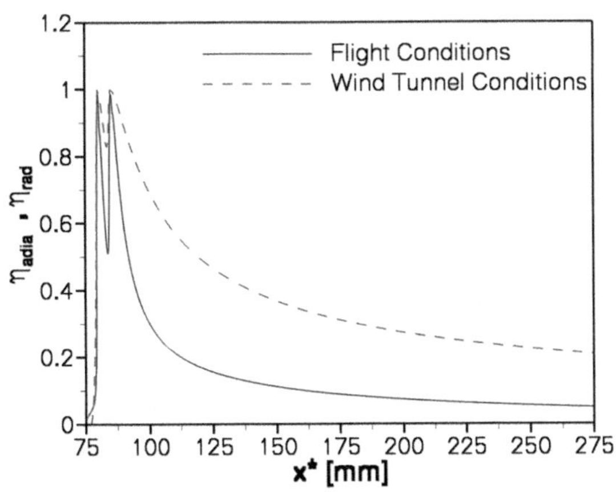

Fig. 12 Downstream evolution of the cooling effectiveness η for a case at wind tunnel conditions (WTC, adiabatic) and for case $\mathscr{A}$ at flight conditions (FC, radiation adiabatic)

2.4 Blowing Through Holes

Here the cool air is blown through four successive rows of holes into the boundary layer, with the position of the holes somewhat further downstream of the leading edge ($x_c = 1.998$). The rows of holes are staggered in the discussed cases ($\mathscr{C}$ – $\mathscr{E}$) because in [8] it was shown that the cooling effectiveness is higher than in the aligned configuration. Thus we concentrate on the influence of the spanwise (s_z) and streamwise spacing (s_x) of the holes onto the cooling effect. Note that the maximum blowing ratio $(\rho v)_{c,max}$ and the cooling gas temperature ($T^{\star}_{c,core} = 293$ K) are in all three cases the same.

Table 2 Parameters of the hole configurations for the case at $M_e = 6.8$

case	$(\rho v)_{c,max}$	hole diameter d	spanwise spacing s_z	streamwise spacing s_x
$\mathscr{C}$	0.125	$0.06 \approx \delta$	0.12	0.12
$\mathscr{D}$	0.125	0.06	0.12	0.18
$\mathscr{E}$	0.125	0.06	0.18	0.12

In Fig. 13 the wall temperature for these cases is shown. First we see from cases $\mathscr{C}$ and $\mathscr{D}$ that the average wall temperature is even lower in the case of larger streamwise spacing. Here, with only 67% coolant-gas mass flow per streamwise unit of case $\mathscr{C}$ — taken at the fourth row of case $\mathscr{D}$ — the counter-rotating vortex pairs (CVPs) induced by each cooling jet have more space to fully form unlike case $\mathscr{C}$, and therefore can push the coolant gas effusing from the next, staggered row more effectively down resulting in a more homogeneous coolant film. The CVPs are along the jet trajectory and have a rotation sense such that the coolant gas is transported away from the wall in the streamwise hole center line. The spanwise spacing (case $\mathscr{E}$) has an even stronger effect. Here the wall-temperature decrease is indeed much lower because of the lower cooling-gas input per spanwise unit. It can clearly be seen that the CVPs remain close to the hole center line downstream, and with s_z too large ($> 2\delta$), the CVPs from the staggered holes do not (positively) interfere resulting in a dramatic cooling loss due to the absence of a coolant film. A crosscut of the temperature field at the hole centers of the second row for cases $\mathscr{C}$ and $\mathscr{E}$ is shown in Fig. 14. In case $\mathscr{C}$ (Fig. 14 right) the boundary-layer thickness is larger because of the higher mass flow per spanwise unit despite the layer is "cooler" . The vortices emanating from the first hole row lie near $z = 0$. In case $\mathscr{C}$ the one vortex lies closer to the second hole and pushes the cool gas effusing from the second hole to the wall. The interaction of the effusion jet and the vortices from upstream is much lower in case $\mathscr{E}$ (Fig. 14 left). More hot gas can reach the near-wall region with too large a spanwise hole spacing.

The downstream evolution of the averaged wall temperature $\overline{T}_w$ is illustrated in Fig. 15. From this figure it can be seen that there is hardly a difference between the cases with different streamwise spacings (cases $\mathscr{C}$ and $\mathscr{D}$), where $\mathscr{C}$ needs 50% more coolant-gas mass flow per streamwise unit. There is an upstream effect of the effusion cooling which is strongest in case $\mathscr{C}$. In front of the first row a small recircu-

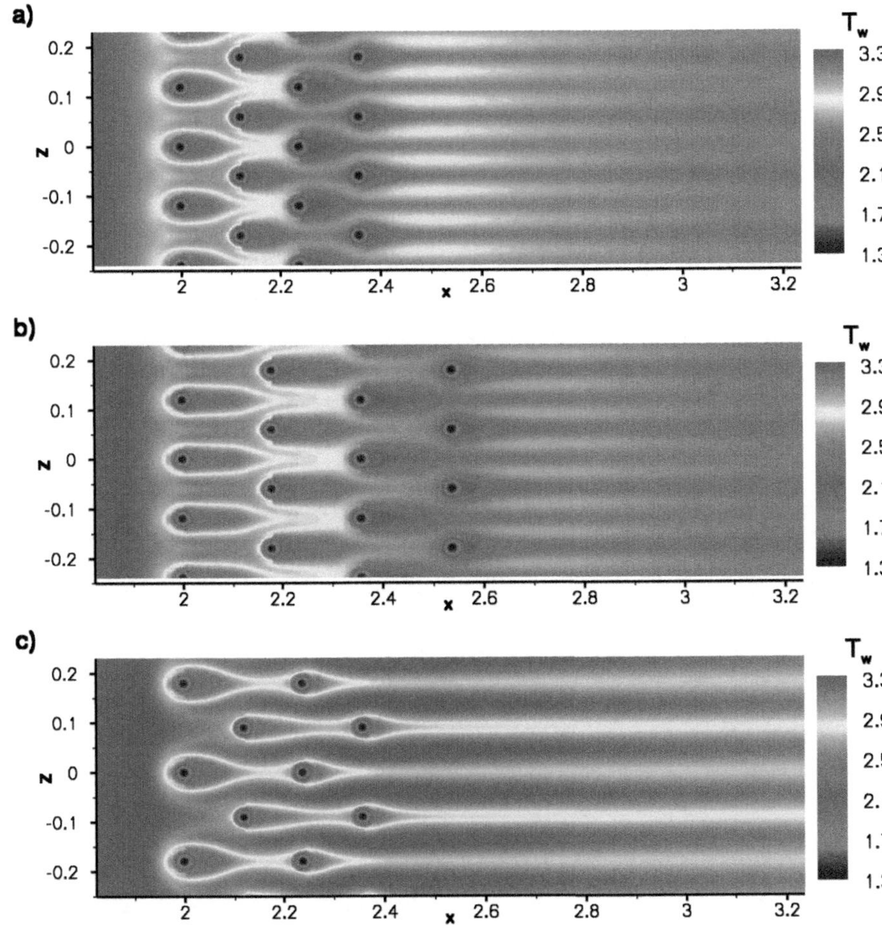

Fig. 13 Wall temperature for steady blowing through four rows of holes into a radiation-adiabatic flat-plate boundary layer at Mach 6.8. **a** case $\mathscr{C}$ with $s_x = s_z = 2\delta$, **b** case $\mathscr{D}$ with $s_x = 3\delta$ and $s_z = 2\delta$, and **c** case $\mathscr{E}$ with $s_x = 2\delta$ and $s_z = 3\delta$.

lation region forms transporting cold gas slightly upstream. Case $\mathscr{E}$ with persistent wall-temperature streaks causes the wall at $x = 3$ to be about 115 K in average hotter than the other cases, that reach there their maximum cooling of roughly 210 K. Note that half the streaks show almost uncooled temperature, a pattern that however will be somewhat averaged out with inclusion of (lateral) heat conduction in the wall structure in a gas-wall interaction computational model.

For case $\mathscr{C}$ the vortical structures are visualized via the λ_2-criterion in Fig. 16. At the hole array strong vortical structures are generated and interact with each other. The CVPs of the successive rows are pushed away from the wall when they reach the next row. We note that they have a rotation sense and appearance like

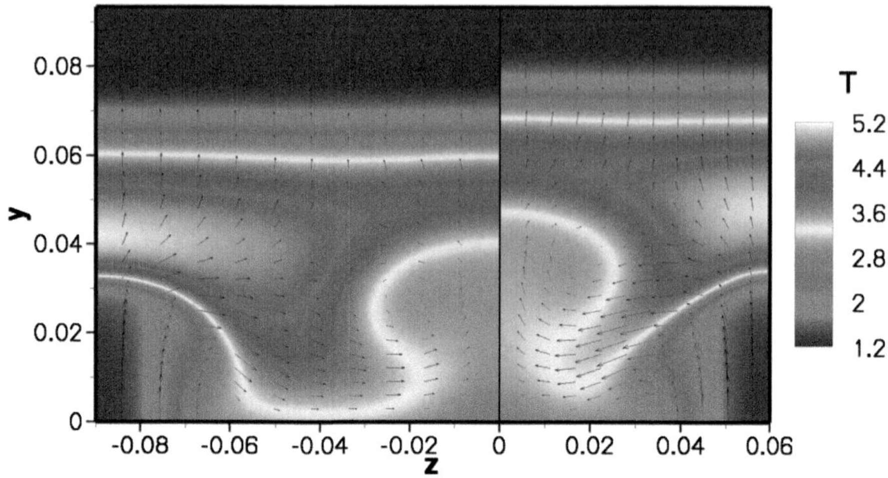

Fig. 14 Crosscut of the temperature field and velocity vectors at the second row of holes ($x = 2.11$) for case $\mathscr{E}$ with the large spanwise spacing (left) and case $\mathscr{C}$ with small spanwise spacing (right)

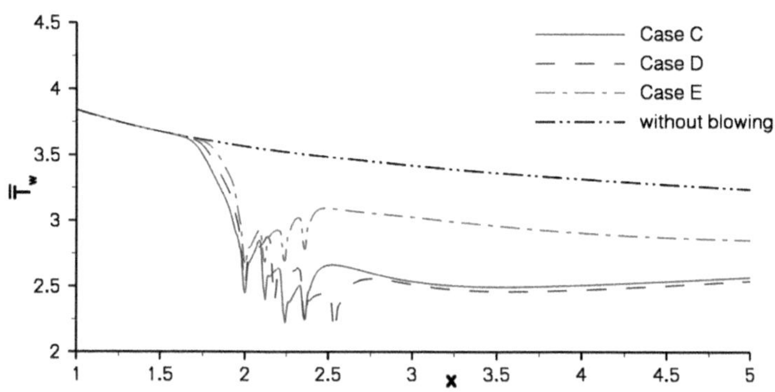

Fig. 15 Spanwise averaged temperature at the wall $\overline{T}_w$ for the cases with four staggered rows of holes and different spanwise and streamwise hole spacing ($\mathscr{C}$, $\mathscr{D}$, $\mathscr{E}$) and for the case without blowing

traveling Λ-vortices during K-type laminar-turbulent transition and thus can easily induce transition with unsteady disturbance parts. Downstream of the last row parallel longitudinal vortex structures, pairwise counter-rotating, are present that decay downstream consistent with Fig. 13a. In front of the first hole row horseshoe vortices exist, like for solid obstacles in the boundary-layer flow, whose rotation sense is opposite to the CVPs and to the neck vortices that form at each hole edge by blowing. All vortex structures lie in a subsonic region at the used blowing rate. We point out that with even lower blowing self-excited unsteadiness and transition to turbulence is found in a subsonic boundary layer. Thus the laminar high-speed flow is more stable with respect to localized three-dimensional blowing, also caused by a smaller wall shear.

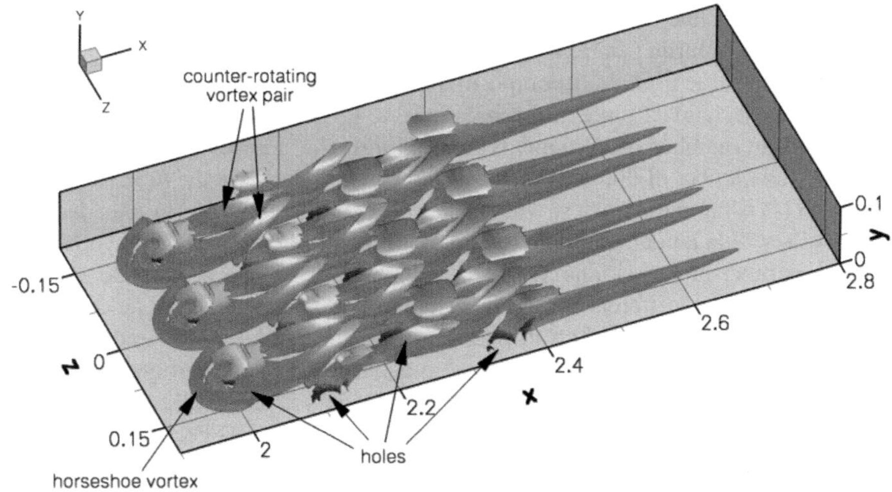

Fig. 16 Visualization of the vortical structures with λ_2 for four staggered rows of holes (case $\mathscr{C}$)

3 Computational Aspects and Outlook

The incompressible code N3D has been optimized in the frame of a Teraflop Workbench project for the NEC SX-8 with its 8 CPUs per node (see also [10, 11]). It primarily computes in Fourier space because of the Fourier-spectral ansatz in spanwise direction that decouples the three-dimensional Poisson-type equations in $(K+1)$ independent two-dimensional problems, where K is the maximum Fourier-mode number. The code is parallelized using OpenMP (intranodal) and MPI (typically internodal), and a speed-up of a factor of about 2 has been achieved by improving the communication and employing optimized FFTs within the Workbench project.

Each node works on 8 spectral components in the optimal case for the Poisson-equation part. The nonlinear convective terms are computed pseudospectrally, i.e in physical space, and thus a transformation to physical space and back is performed using optimized FFTs. To avoid aliasing by the nonlinear generation of higher modes only $2/3$ of the gained modes are used: Using $K = 10$ modes for example means adding 6 modes with zero values, transformation to 32 points ($2^{K_{exp}}$, $K_{exp} = 5$) in physical space, computation of the nonlinear terms, and back transformation keeping only 10 modes of the 16. The MPI parallelization is then done for blocks in the chordwise x-direction, without necessity for computing derivatives, and within each node the CPUs do slices of the wall-normal y-direction in parallel. Because the FFTs are based on powers of 2, optimal values of K exist that both exploit relatively well the computer architecture and minimize the FFT work. Such pairs are (K, K_{exp}): (10, 5), (21, 6), (42, 7), (79, 7), ... , where in the last given case theoretically no idling CPUs occur. The code typically needs 0.9 μs per grid point and time step and is not yet adapted to curvilinear grids.

The compressible code NS3D primarily computes in physical space and is currently subject to further optimization despite being already quite fast compared to other codes. A true domain decomposition (in the $x-y$ plane) is implemented, i.e. derivatives by compact finite differences have to be computed over domain boundaries, and the number of domains typically limits the number of used nodes. Increasing the number of domains then can significantly decrease the turnaround time for a job by using many nodes, whereas for N3D in its present form K is decisive. The CPUs within a node compute longitudinal cuts of the flow field by working on a given number of z-positions. Optimal combinations of spanwise-mode numbers and domains exist, different for symmetric and non-symmetric flow fields. A grid transformation is embedded so that arbitrary bodies can be considered. Thus the code is more flexible coping also with aero-acoustic problems and typically needs 1.8 μs per grid point and time step. However it needs much smaller time steps — about one order of magnitude — for subsonic flows than N3D because of the time step limit governed by the transport of fast sound waves. For a subsonic laminar-turbulent transition problem as discussed in the LFC case above the compressible code would need about 20 times as long in CPU time.

Until now our typical supercomputing data, including simulations in the group of U. Rist at IAG, were for:

- N3D, laminar flow control of a swept-wing flow by suction or active control, turbulent separation control using inclined slot blowing, control of laminar separation bubbles, mechanisms/control of boundary-layer transition: 1 billion grid points, 0.4 TB RAM (0.4 KB/point), 34 nodes, 4 GFlop/s per CPU, 1.1 TFlop/s, 100 h wall time.
- NS3D, mechanisms and control of shear-layer noise, hypersonic transitional boundary-layer flow on plates/cones: 140 million grid points, 0.162 TB RAM (1.1 KB/point), 16 nodes, 5.4 GFlop/s per CPU, 0.7 TFlop/s, 46 h wall time.

Our plans for the near future are:

- N3D, more transition/turbulence/active control in two- and three-dimensional base flows, higher Reynolds numbers, larger domains for wings, possibly more complex geometry — needing 5 billion grid points and about 10 GFlop/s per CPU.
- NS3D, disturbance receptivity/transition/control in three-dimensional base flows for high subsonic Mach numbers and supersonic Mach numbers with shock layers, complete flow around wing profiles with disturbance feed back, high-temperature effects in hypersonic shear flows including ducts in the wall — needing 20 billion grid points, more nodes, and for a compressible bi-global stability eigenvalue solver 1 TB RAM per node.

For any (growing) up-to-date problem the user wants a non-growing turnaround time, and a computer that is as stable as the NECs currently used at HLRS.

Acknowledgements The authors gratefully acknowledge the financial support of the Deutsche Forschungsgemeinschaft (DFG), the Helmholtz-Gemeinschaft Deutscher Forschungszentren (HGF), and the support of the High Performance Computing Center Stuttgart (HLRS) within the Teraflop project.

References

1. Babucke, A., Linn, J., Kloker, M.J., Rist, U.: Direct numerical simulation of shear flow phenomena on parallel vector computers. In: M. Resch et al. (eds.) High Performance Computing on Vector Systems 2005, pp. 229–247, Springer (2006)
2. Bippes, H.: Basic experiments on transition in three-dimensional boundary layers dominated by crossflow instability. Progress in Aerospace Sciences **35**, 363–412 (1999)
3. Bonfigli, G., Kloker, M.J.: Secondary instability of crossflow vortices: validation of the stability theory by direct numerical simulation. J. Fluid Mech. **583**, 229–272 (2007)
4. Friederich, T.: Active control of the crossflow secondary instability in a 3-d boundary layer using steady blowing and suction. Master's thesis, Institut für Aerodynamik und Gasdynamik, Universität Stuttgart (2007)
5. Friederich, T., Kloker, M.J.: Localized blowing and suction for direct control of the crossflow secondary instability. AIAA-2008-4394, Seattle (2008)
6. Hirschel, E.H.: Basics of Aerothermodynamics. Springer (2004)
7. Koch, W., Bertolotti, F. P., Stolte, A. and Hein, S.: Nonlinear equilibrium solutions in a three-dimensional boundary layer and their secondary instability. J. Fluid Mech. **406**, 131–174 (2000)
8. Linn, J., Kloker, M.J.: Numerical Investigations of Film Cooling. In: A. Gülhan (ed.) RESPACE — Key Technologies for Resuable Space Systems, NNFM 98, pp. 151–169, Springer (2008)
9. Maucher, U., Rist, U., Wagner. S.: A method for the identification of high-frequency oscillations in unsteady flows. ZAMM **77**, 209–210, Suppl. 1 (1997)
10. Messing, R., Rist, U., Svensson, F.: Control of turbulent boundary-layer flow using slot actuators. In: M. Resch et al. (eds.) High Performance Computing on Vector Systems 2006, pp. 91–103, Springer (2007)
11. Messing, R., Kloker, M.J.: Smart suction — an advanced concept for laminar flow control of three-dimensional boundary layers. In: M. Resch et al. (eds.) High Performance Computing on Vector Systems 2007, Springer (2008)
12. Saric, W., Reed, H.L., White, E.B.: Stability and transition of three-dimensional boundary layers, Annu. Rev. Fluid Mech. **35**, 413–440 (2003)
13. Wassermann, P., Kloker, M.J.: Mechanisms and passive control of crossflow-vortex-induced transition in a three-dimensional boundary layer, J. Fluid Mech. **456**, 49–84 (2002)
14. White, E. B., Saric, W. S.: Secondary instability of crossflow vortices, J. Fluid Mech. **525**, 275–308 (2005)

Fluid-Structure Interaction in Turbine Simulation

Felix Lippold and Albert Ruprecht

Abstract In this article, two examples of fluid-structure interaction (FSI) are examined. Furthermore, the issues of turbulence and the impact of turbulence models on the accuracy of Karman vortices is evaluated. For this purpose, an adaptive turbulence model is described, introduced and validated. For the FSI simulations, a partitioned scheme is introduced and implemented. Since the main application considered here is hydraulic machinery, the issue of added mass effects and unstable coupling is addressed and discussed. The scheme is validated with a benchmark application established recently. Finally, the first results obtained for the FSI coupling of a tidal current turbine runner blade under fluid loads are described. For the simulations, the established in-house CFD-Code FENFLOSS is used. For the coupling, the commercial software MpCCI is applied. ABAQUS and CalculiX are used for the solution of the structural part. Simulations are performed on a cluster and a NEC SX-8 vector computer.

1 Introduction

In many fields of engineering complex vortex dominated flows play an important role. Often the flow behaviour gets unstable and a complex unsteady vortex movement occurs. These vortices lead to a pulsating pressure field and consequently also to unsteady loads on the structure.

In case of slim structures the unsteady loading can lead to a deformation of the structure. A relatively large deformation of the structure on the other hand can again influence the flow behaviour. For an accurate prediction of the flow behaviour, at first, the unsteady vortex motion has to be simulated very accurately. For example by using an inappropriate turbulence model one can completely suppress unsteady

Felix Lippold · Albert Ruprecht
Institut für Strömungsmechanik und Hydraulische Strömungsmaschinen, Universität Stuttgart, Pfaffenwaldring 10, 70550 Stuttgart, Germany, e-mail: {lippold, ruprecht}@ihs.uni-stuttgart.de

effects. Here different sophisticated models are analysed for the prediction of an unsteady vortex movement. The used application is the vortex street behind a straight trailing edge of a runner blade in a water turbine. This is an example that requires a sophisticated and accurate model to predict the unsteady vortices.

Further, in a second step, a coupled solution of the fluid and of the resulting structural behaviour is necessary. For this fluid-structure interaction the coupling of a flow simulation code and a structure simulation code is carried out. As an example the flow through a tidal current turbine is taken into account. This type of turbine is characterized by rather slim and long turbine blades, where the deformation cannot be neglected.

In this paper an overview over the used turbulence models and its performance is given. Also the fluid-structure coupling is described in detail. Both, the turbulence model and the coupling method are evaluated with simplified models and subsequently used to analyse a tidal current turbine runner blade.

2 Basic Equations

Incompressible, turbulent flows are described by the continuity equation for constant fluid density and the Reynolds-averaged Navier-Stokes equations. If the fluid domain is changing, e.g. in fluid-structure interaction, the mesh movement has to be accounted for. This yields the Arbitrary-Lagrange-Euler (ALE) formulation. Furthermore, turbulent flow plays an important role in engineering applications. Nevertheless, a direct simulation of turbulent fluctuations is not possible for complex problems, today. Hence, the influence of the turbulent fluctuations on the mean flow is modelled by an additional viscosity. Two-equation models solve two additional transport equations. They are widely used today even though they have certain drawbacks. But, due to sophisticated enhancements to the standard models a wide range of flow phenomena can be represented with acceptable accuracy.

An additional topic in fluid-structure interaction is the representation of the structure. In hydraulic machinery the added fluid mass has a great influence on the vibrational behaviour. The transformation of the fluid pressure to the structural surface models this additional mass. Usually, a linear material behaviour with small deformations may be assumed.

Flow equations

In order to simulate the flow of an incompressible fluid the momentum equations and mass conservation equation are derived in an infinitesimal control volume. Including the turbulent fluctuations yields the incompressible Reynolds-averaged Navier-Stokes equations [6]. Considering the mesh deformation introduces the velocity of the grid nodes U_{G} and results in the Arbitrary-Lagrange-Euler (ALE) formulation [4, 11]

$$\frac{\partial U_i}{\partial x_i} = 0\,, \tag{1}$$

$$\left.\frac{\partial U_i}{\partial t}\right|_{\mathscr{X}} + \underbrace{(U_j - U_{\mathrm{G}})}_{=\hat{U}_j}\frac{\partial U_i}{\partial x_j} + \frac{1}{\rho}\frac{\partial P}{\partial x_i}$$

$$+\frac{\partial}{\partial x_j}\left[\nu\left(\frac{\partial U_i}{\partial x_j} + \frac{\partial U_j}{\partial x_i}\right) - \overline{u_i' u_j'}\right] = f_i\,. \tag{2}$$

The components of the flow velocity are U_i and the pressure is P. On the right hand side there are loads due to rotational and body forces f_i. In this case we use an ALE-formulation of the Navier-Stokes equations, hence, the time derivatives are computed in the ALE frame of reference $\mathscr{X}$. In case of a non-moving mesh, $U_{\mathrm{G}} = 0$, Eq. (2) complies with the regular Euler formulation of the Navier-Stokes equations.

Turbulence Modelling

Usually, Reynolds Stresses are modelled following Boussinesq's eddy viscosity principle. To model the resulting turbulent viscosity, for many engineering problems k-ε and k-ω-models are combined with logarithmic wall functions. Alternatively, Low-Reynolds formulations are applied. For both models and their enhanced variations two additional transport equations have to be solved. The turbulent viscosity ν_{t} is computed from the turbulent kinetic energy k and its dissipation ratio ε or the turbulent vorticity ω, respectively. Basically, the k-equation stays the same for both models and reads for the standard versions

$$\frac{\partial k}{\partial t} + \hat{U}_i\frac{\partial k}{\partial x_i} - \frac{\partial}{\partial x_i}\left[\left(\nu + \frac{\nu_t}{\sigma_k}\right)\frac{\partial k}{\partial x_i}\right] = G - \varepsilon\,. \tag{3}$$

The equations for the dissipation ratio and the turbulent vorticity read

$$\frac{\partial \varepsilon}{\partial t} + \hat{U}_i\frac{\partial \varepsilon}{\partial x_i} - \frac{\partial}{\partial x_i}\left[\left(\nu + \frac{\nu_t}{\sigma_\varepsilon}\right)\frac{\partial \varepsilon}{\partial x_i}\right] = c_{1\varepsilon}\frac{\varepsilon}{k}G - c_{2\varepsilon}\frac{\varepsilon^2}{k} \tag{4}$$

$$\frac{\partial \omega}{\partial t} + \hat{U}_i\frac{\partial \omega}{\partial x_i} - \frac{\partial}{\partial x_j}\left[\left(\nu + \frac{\varepsilon}{\sigma_\omega}\right)\frac{\partial \omega}{\partial x_j}\right] = \alpha\frac{\omega}{k}R_{ij}\frac{\partial u_i}{\partial x_j} - \beta\omega^2\,. \tag{5}$$

The model constants may vary depending on the model. A good reference is given in [2].

Unsteady vortex simulations usually require more sophisticated turbulence models than those shown above. For this reason there are certain enhancements available. Most of them only add further expressions to the transport equations which minimises the additional computational effort. In this paper we use the Standard k-ε model [14] for most computations. In case of the draft tube vortex the results are compared to those obtained with the SST-Model [17] and an enhanced k-ε (Kim-

Chen) [3, 10]. Both models are based on versions of the k-ε and k-ω model. Hence, the computational effort is comparable to the Standard models.

Very Large Eddy Simulation (VLES)

"Real" Large Eddy Simulation (LES) from the turbulence research point of view require an enormous computational effort since all anisotropic turbulence structures have to be resolved in the computation and only the smallest isotropic scales are modeled. Consequently this method can not be applied for industrial problems today. Today's calculations of flows of practical relevance (characterized by complex geometry and high Reynolds number) are usually based on the Reynolds-averaged Navier-Stokes (RANS) equations. This means that the influence of the complete turbulence behavior is expressed by means of an appropriate turbulence model. To find a turbulence model, which is able to capture a wide range of complex flow effects quite accurate is impossible. Especially for unsteady flow behavior this method often leads to rather poor results. The RANS and LES approach can schematically be seen in Fig. 1, where a typical turbulent spectrum and its division in resolved and modeled parts is shown.

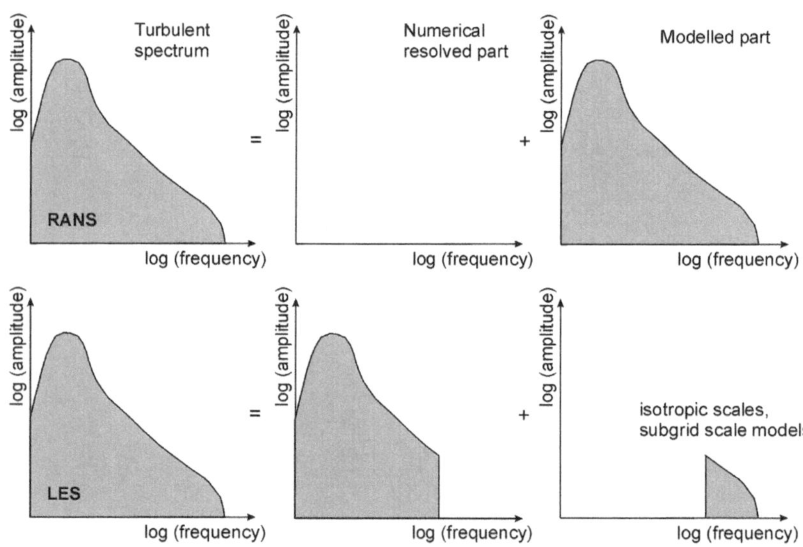

Fig. 1 Modelling approach for RANS and LES

This established approach leads to quite promising results, especially for unsteady vortex motion [10]. Contrary to URANS there is a requirement to the applied turbulence model, so that it can distinguish between resolved unsteady motion and not resolved turbulent motion which must be included in the model. The VLES approach is a compromise between RANS and LES. The very large eddies are resolved and the smaller structures (still anisotropic) are modelled, this is schematically shown in Fig. 2. VLES is also found in the literature under different other names:

- Semi-Deterministic Simulation (SDS),
- Coherent Structure Capturing (CSC),
- Detached Eddy Simulation (DES),
- Hybrid RANS/LES,
- Limited Numerical Scales (LNS).

- Hybrid RANS/LES,
- Limited Numerical Scales (LNS).

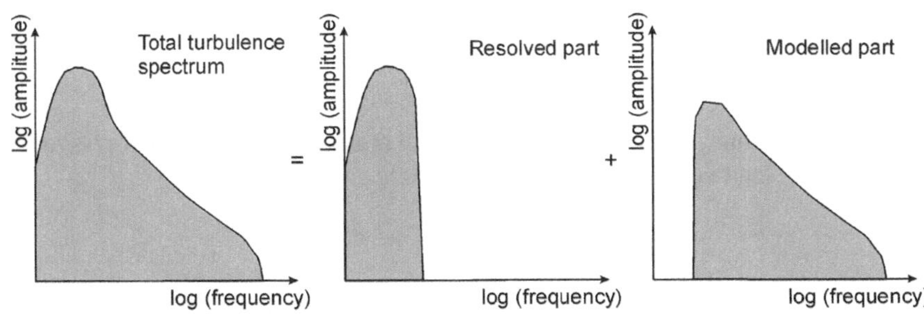

Fig. 2 Turbulent spectrum resolved and modelled by VLES

Adaptive Turbulence Model

Classical turbulence models, which are usually applied in engineering flow predictions, contain the whole turbulent spectrum. They usually show a too viscous behavior and very often damp out unsteady motion. As discussed above the turbulence model for the VLES has to distinguish between the resolved and unresolved part of the turbulent spectrum (Fig. 2). Therefore an adaptive model is used, which adjusts its behavior according to the approach (schematically shown in Fig. 3). This means that this model can be applied for all approaches.

The advantage of the adaptive model is, that with increasing computer power the resolved part of the turbulent spectrum increases and the modeled part decreases, consequently the accuracy of the calculations improves.

There are several filtering techniques in the literature. Here a filtering technique similar to [24] is applied. The basis of this adaptive model is the modified k-ε model of Chen and Kim [3]. This model has been chosen because it is quite simple and its results are much better, especially for unsteady flows, than the ones obtained with the standard k-ε model. By using the filtering technique the turbulent structures

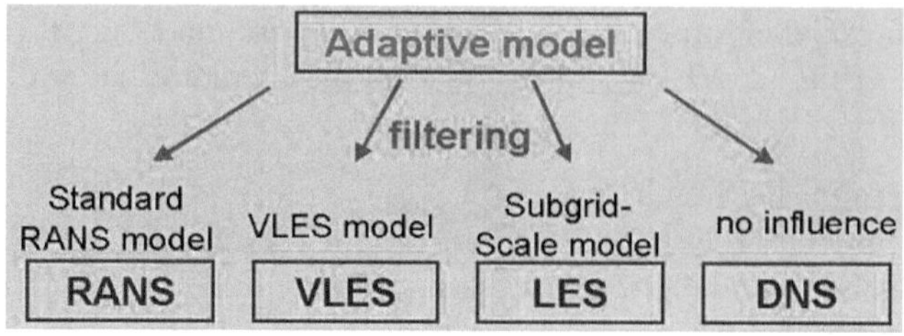

Fig. 3 Adjustment for adaptive model

are resolved or modelled depending on mesh and flow characteristics. Furthermore, there are no more transport equations to be solved than for a regular turbulence model. Hence, the only additional effort is for the filtering, which is negligible.

Structural Equations

The discretised linear structural equations with mass, damping and stiffness matrices M, D, and K, load vector f and displacements u can be written as

$$M\ddot{u} + D\dot{u} + Ku = f \,, \tag{6}$$

see Zienkiewicz [25]. For fluid-structure coupled systems the load vector f represents the fluid loads due to surface pressure and inertia.

3 Fluid-Structure Coupling

Seen from the physical point of view fluid-structure interaction is a two field problem. But numerically, the problem has three fields, where the third field is the fluid grid that has to be updated after every deformation step of the structure to propagate the movement of the fluid boundaries, the wetted structure, into the flow domain.

The solution of the numerical problem can be arranged in a monolithic scheme, which solves the structural and flow equations in one step. This method is applied for strongly coupled problems, e.g. for modal analyses. Another scheme, suitable for loosely and intermediately coupled simulations, is the separated solution of flow and structural equations with two independent codes and models. In the separated scheme well proven codes and models can be used. This is the reason why for engineering problems usually the second approach is employed. However, data has to be exchanged between the codes, including the interpolation between the two computational meshes. In order to account for the coupling and to avoid unstable simulations, some coupling schemes have been developed, e.g. see Farhat et. al. [5, 18].

All schemes shown in Fig. 4 exchange data at each time-step. In the first scheme, a sequentially staggered scheme, one code waits until the other one is done with the time-step. Hence, one code always uses data of the remote code's previous step for the time-step integration. This gives the scheme explicit character. The second explicit scheme is a parallel staggered one. Here, both codes do the time-step integration simultaneously. If the solution requires approximately the same real time in both codes, this scheme will reduce waiting times.

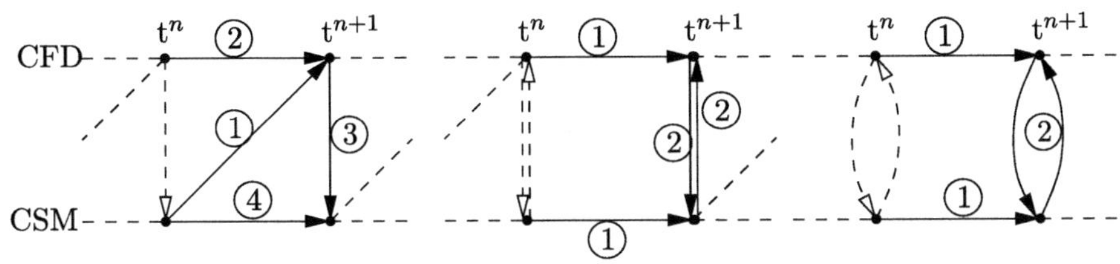

Fig. 4 Partitioned schemes for loose and strong coupling

Explicit coupling schemes will exhibit instabilities if the moving fluid mass is high compared to the structural mass [7]. The high density of water leads to a high added fluid mass in hydraulic machinery applications. These strongly problems require an implicit coupling scheme. Due to the iteration between the codes these schemes increase the computational effort.

In this paper the explicit schemes are applied for quasi-static analyses of coupled problems. For time dependent problems an implicit scheme is used. Furthermore, the difference between the two explicit schemes will be examined.

4 Efficient Moving Mesh Scheme

The mesh update method applied in the computations in this paper uses an interpolation between the nodal distance between moving and fixed boundaries to compute the new nodal position after a displacement step of the moving boundary. The simplest approach is to use a linear interpolation value $0 \leq \kappa \leq 1$. Here we use a modification of the parameter $\kappa = \frac{|s|}{|r|+|s|}$ proposed by Kjellgren and Hyvärinen [13]

$$\tilde{\kappa} = \begin{cases} 0, & \kappa < \delta \\ \frac{1}{2}\cos\left(\left(1 - \frac{\kappa-\delta}{1-2\delta}\right)\pi\right) + 1, & \delta \leq \kappa \leq 1-\delta \\ 1, & 1-\delta < \kappa \leq 1 \end{cases} . \tag{7}$$

This parameter is found from the nearest distance r to the moving boundary and the distance to the fixed boundary s in the opposite direction. To use this approach for parallel computations the boundaries have to be available on all processors. Since a graph based domain decomposition is used here this is not implicitly given. Hence, the boundary exchange has to be implemented, additionally. Usually the number of boundary nodes is considerably small, i.e. the additional communication time and overhead is negligible. The algorithm is discussed and compared with other methods regarding computational performance, parallelisation and robustness in [15]. Compared to methods based on differential equations this algebraic method does not need to solve of a set of discretised equations at each grid-update step. This saves a considerable amount of computing time. Furthermore, in our applications usually, the structure vibrates around a certain starting configuration. An interpolation method, as presented above, guarantees the conservation of the original mesh whenever the structure is in its original shape.

5 Coupled Codes and Architecture

All simulations shown in this paper require a considerable amount of computing power. In case of the vortex simulations a fine mesh is necessary to resolve the high gradients in the flow. Furthermore, several periods of the periodic vortex rope have to be computed. For the fluid-structure simulations the CFD part is computationally expensive compared to the structural simulation. Especially for dynamic FSI simulations a great number of steps has to be solved. Hence, the CFD runs will all be performed on an HPC system, NEC SX-8. This means to verify the performance of the CFD-code FENFLOSS as well as to install the coupling on the vector system.

5.1 FENFLOSS — CFD on Vector Systems

The flow simulation (CFD) code FENFLOSS[1], developped at the IHS, is an unsteady Navier-Stokes solver based on the finite element method. A stabilized Petrov-Galerkin Finite Element approach [25, 9] is applied for the discretization of Eqs. (1) and (2). The nonlinear system is solved by a fixed-point iteration scheme. In each nonlinear iteration loop the equations are linearised and solved by an iterative BICGStab(2) solver [22, 23], Fig. 5. The three velocity components can be solved coupled or decoupled followed by a modified UZAWA pressure correction, as described by Ruprecht [19]. Working on parallel architectures, MPI is applied in the preconditioner, the matrix-vector and scalar products [16]. FENFLOSS supports all unstructured grids with hexahedral (3D) or quadrilateral (2D) elements.

Furthermore, FENFLOSS provides a programming interface (API), which allows to implement user-defined subroutines. These will be called at certain points in the solver loops. The user-subroutines are implemented by a shared object library that is loaded on demand during the initialisation of the CFD solver. This guarantees a very flexible method to enhance the FENFLOSS solver kernel or even exchange data to external programs.

---

[1] **F**inite **E**lement based **N**umerical **FL**ow **S**imulation **S**ystem

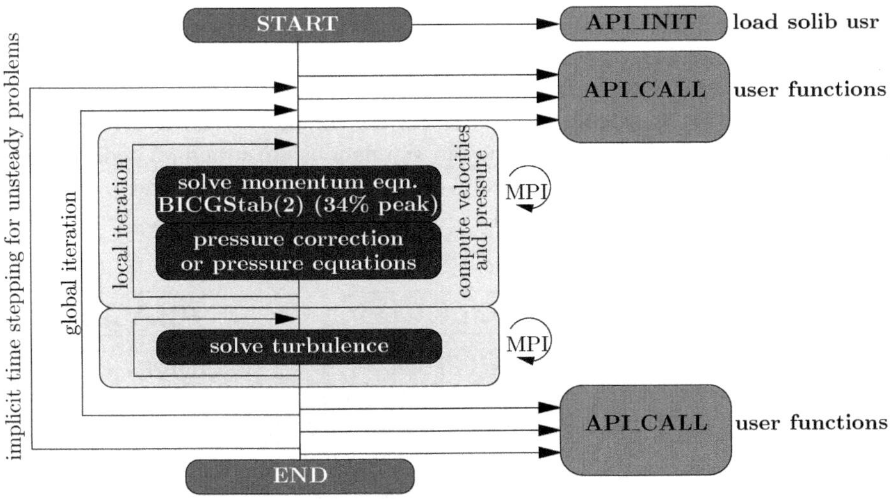

Fig. 5 FENFLOSS solver scheme — including user-function calls

FENFLOSS has been successfully optimised for vector architectures [1]. Despite the indirect addressing the matrix vector product in the linear solver reaches a ratio of almost 50% of the peak performance (16 GFlop/s) on a SX-8 vector CPU for vector lengths near to the optimum of 256. The whole solver still has a relative performance of nearly 35% and the total application, including start-up and I/O gets close to 30%.

5.2 FSI Coupling with FENFLOSS on SX-8

For fluid-structure simulations the code coupling software MpCCI[2], developped at the Fraunhofer Institute SCAI, is applied. It provides a coupling manager that controls the simulation codes and interpolates the data to the different meshes. To communicate data with the codes MpCCI needs a special code adapter.

The MpCCI-adapter for FENFLOSS is available as shared object library. It has access to the pressure solution in FENFLOSS and the geometry data. For the explicit coupling scheme, see Sect. 3, data is transferred once at the beginning of the time-step. A data transfer sends the current pressure values and receives the deflections on the coupling region. If the coordinates of the coupling surface have changed the mesh is updated automatically. For implicit schemes an additional transfer is made in the global iteration loop.

In this paper we use the commercial structural simulation tool ABAQUS(C) and, for the implicit coupling, the open source solver CalculiX, developped at MTU Aero Engines, Munich. Both codes use an implicit Newmark type time-stepping scheme.

[2] Mesh based parallel Code Coupling Interface(C)

For implicitly coupled simulations additional data is exchanged within the inner iteration of CalculiX usually used for non-linear analysis.

The visualisation software COVISE is used for post-processing of the CFD results. Figure 6 shows the complete set-up for the simulation. ABAQUS stands for the structural simulation code. During our simulations the structural code will run on a single cluster CPU that is allowed to connect to the SX-8 front end to exchange data with the CFD solver via MpCCI.

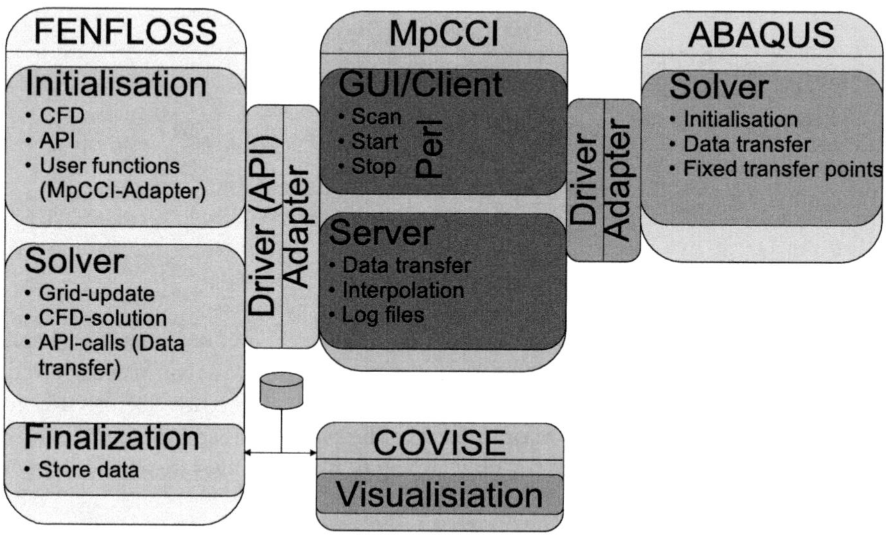

Fig. 6 FENFLOSS coupled with MpCCI and ABAQUS

For the installation of the MpCCI-coupled version of FENFLOSS on the NEC SX-8 vector system we have to make some modifications. The compiler installed with Super-Unix 15.1 on the SX-8 does not support the concept of shared object libraries. Hence, we have to give up the flexibility this approach gives us on regular Linux systems and link the user-defined functions statically to the FENFLOSS kernel. This means to have a special version that only runs with MpCCI. If something in the adapter is changed, the whole program has to be rebuilt.

6 von Karman Vortex Street

In this case study, Karman vortices behind an axial turbine runner blade are examined. The runner has a nominal speed of 1000 rpm, nominal discharge of 0.7 m^3/s and nominal head of 13.2 m. The main diameter and the hub-to-tip ratio are 0.42 m and 0.5, respectively. The boundary conditions used for the 2D flow calculations of an unrolled cylindrical cut next to the hub are computed from the design data above. Since the rotation of the wing will not be considered the velocities are transformed into the rotating system. Furthermore, three different meshes consisting of 100000 (coarse), 230000 (fine) and 290000 (finest) nodes are built to examine the dependency of the solution on the grid. The finest grid uses a non-matching interface between a refined boundary layer grid and a coarser far-field grid. Constraints are introduced by the translational periodicity of the mesh. The quality of the computational results are judged regarding the computed Strouhal number St based on Gongwer's theory [8] introducing an additional virtual thickness

$$\delta_v = 0.0239 \frac{t}{(Re_l)^{\frac{1}{5}}}, \quad l \ldots \text{chord length} \tag{8}$$

to account for the boundary layer.

Following Schlichting [20], the Strouhal number for this application is St = 0.21 referring to the trailing edge thickness (1.5 mm) of the blade. The reference velocity value is the average velocity in the transport direction at the trailing edge excluding the boundary layer and the edge itself.

Computations with the extended k-ε-Model by Kim and Chen show a relative error for computed and theoretical values of about 50%, compared with the frequency according to Gongwer. The VLES model yields an error of 1% for the fine and the finest grid, see Fig. 7.

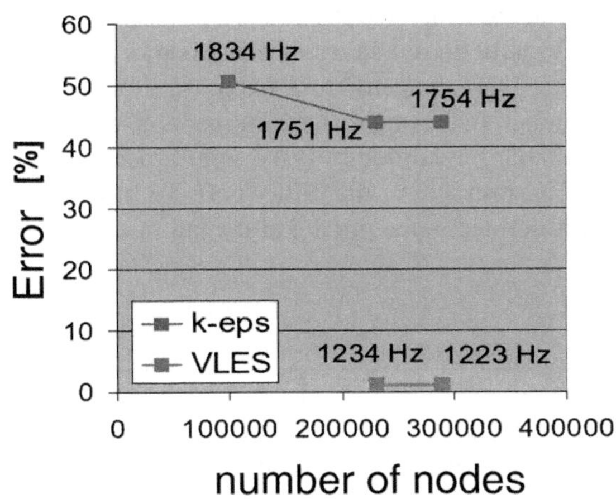

Fig. 7 Relative error of computed and theoretical frequencies

Fig. 8 shows the Karman vortices behind the trailing edge of the blade. The velocities on the pressure (PS) and suction side (SS) of the blade are different. Due to this, the vortices are not absolutely symmetrical, as one would expect for a cylinder.

Based on these results the VLES-model seems to be suitable for the prediction of unsteady vortices. A further detailed analysis of the VLES can be found in [10]. But the resolution of the turbulence effects require a rather fine mesh; already in

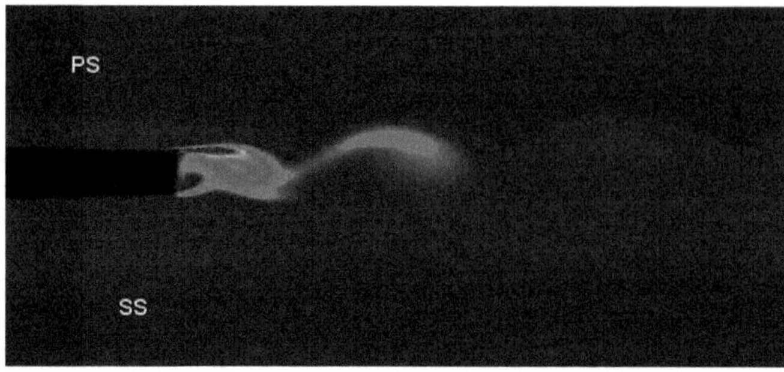

Fig. 8 Dissipation ratio illustrating Karman vortices behind the trailing edge

2D. Hence, to model the turbulent structures more accurately in 3D than regular turbulence models do, one requires a very fine mesh, too. Therefore, already the CFD-analysis without fluid-structure coupling requires a great amount of computational power.

7 FSI Benchmark Application

In order to have a measure of how suitable a fluid-structure coupling method and implementation is, Turek defines a series of benchmarks [21]. It consists of a 2D-channel with an elastic flap connected to a fixed, rigid cylinder. This assembly is asymmetrically positioned in the channel. The upper and lower boundaries are rigid walls with noslip boundary conditions. At the inlet a parabolic (laminar) velocity profile is assumed. The pressure at the outlet is 0. To monitor the deformation a monitoring Point A is defined in the middle of the tail of the flap, Fig. 9.

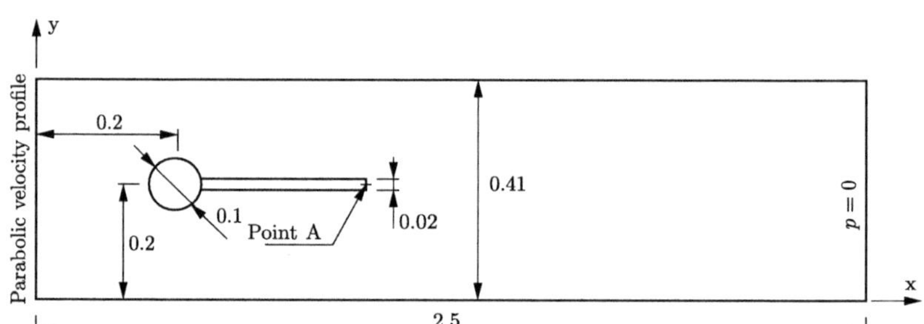

Fig. 9 Benchmark geometry from [21], measures in m

Here we choose the third out of three benchmark configurations, FSI3, for the evaluation of the methods described above. The velocity profile at the inlet is a parabolic profile with a mean velocity $\bar{U} = 2.0\,\mathrm{m/s}$ and the Reynolds Number referring to the cylinder diameter is Re = 200. The results obtained will be compared to those obtained by Turek with a monolithic approach implemented in the public domain code FEATFLOW.

The FSI3 benchmark is supposed to be the most challenging benchmark in the FSI-series. Since the ratio between fluid and structural density ρ_F and ρ_S, respectively, is 1.0, it probably tends to be unstable for explicit couplings. The Young's modulus of the structure is E = 5.6×10^6 N/m^2 and the Poisson ratio 0.4, which corresponds to very flexible rubber-like material. Simulations are performed with a rather fine 2D CFD-mesh with 182,000 nodes. The structural mesh consists of 702 linear rectangular elements. As expected, explicitly coupled simulations show immediate unstable behaviour. The implicit scheme implemented by the MpCCI-coupling of FENFLOSS and CalculiX yields smooth and stable force and deflection trajectories.

Figure 10 shows the forces on the structure and the displacements of the reference point A computed with a time-step of 0.001 s. An FFT-analysis yields a frequency for the vertical forces and movement of 5.53 Hz, which is satisfying compared to 5.3 Hz obtained with FEATFLOW. The difference in the forces is in the same range of 5 to 10%. The fluctuations in the amplitude of the x-displacements are also observed in the reference examples. Even though the structure is exhibits considerably large displacements the simple but efficient moving mesh method presented above still yields a good mesh quality, Fig. 11.

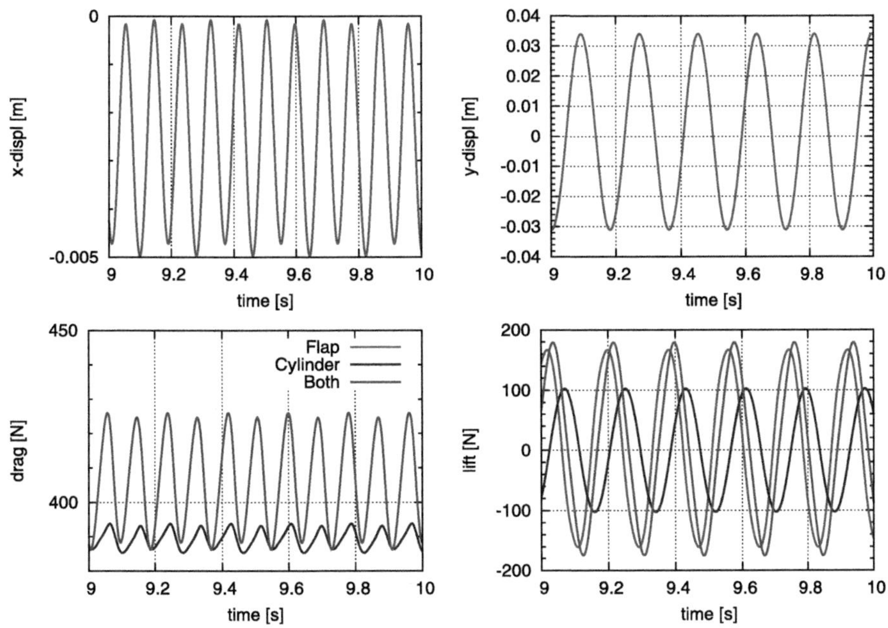

Fig. 10 Displacement and force trajectories obtained for FSI3

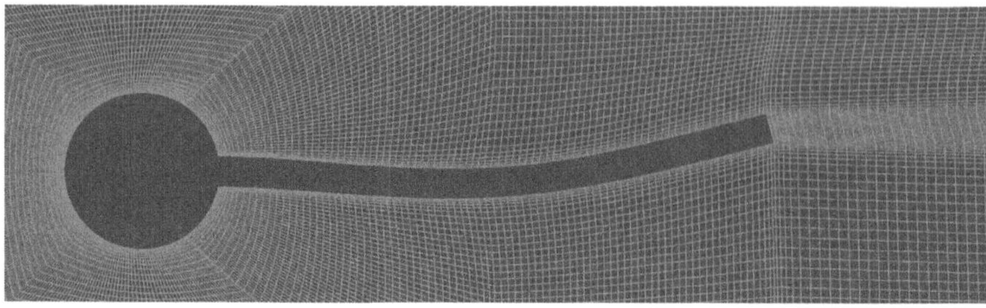

Fig. 11 Mesh around the flap at maximum deflection.

With a relaxation of 0.8 for the displacements and 1.0 for the pressure the computing time on 4 CPUs (3.2 GHz, AMD Opteron) is about 115 h for 10000 time-steps. This is still much faster than the CFX-Ansys-Structural combination described in [12]. The number of CFD-iterations per time-step is about 2 to 3 times the number of iterations needed for a regular CFD-analysis. Figure 12 shows the pretty good convergence behaviour of the coupling. The convergence criteria for the change in displacements is $\Delta y \leq 5 \cdot 10^{-7}$ for each direction.

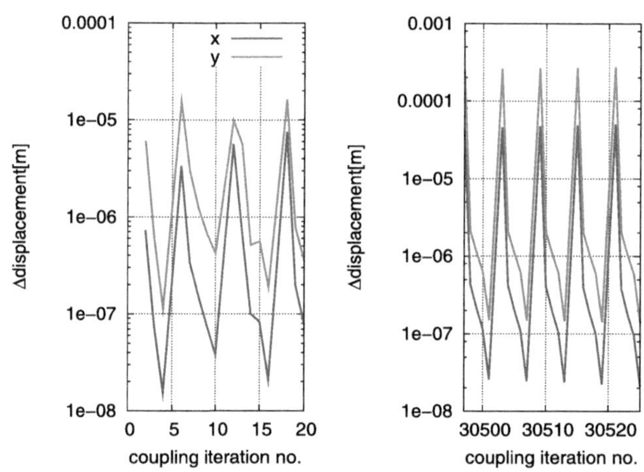

Fig. 12 Convergence behaviour of the coupling

Summarizing, the presented partitioned approach obviously holds the requirements for a reliable fluid-structure interaction simulation for turbine applications. Even though, the results do not perfectly match the reference values obtained by Turek with a monolithic scheme, the absolute differences in force, displacement and frequency values are acceptable. It is clear that the computational effort for the fluid simulation is much more compared to the structural problem due to the finer meshes.

8 Tidal Turbine Runner

Tidal current turbines are similar to wind turbines. They only use the kinetic energy of the flow. As a consequence they have rather long, slim blades. In Fig. 13 a prototype of a tidal current turbine is shown. The diameter of the rotor is 16 m. The flow field to the runner usually is not uniform. Because of a boundary layer

Fig. 13 Computer rendered picture of tidal current turbines

on the bottom and because of the wakes behind the support structure the flow to a turbine blade is unsteady. As a consequence this non-uniform flow field leads to a severe periodic loading on the blades and finally to a noticeable deformation of the blade. To calculate the resulting vibrations of the runner blades is the final task of this project.

After computing a first solution for steady inflow boundary conditions the pressure loads are used to evaluate the static behaviour of one turbine blade. Since the change of the flow due to the deflection is not examined this analysis is called a one-way coupled analysis. This analysis is made to verify the further results. The second step is a two-way coupled analysis, which involves a direct interaction between the flow and the structural solution. Both codes exchange data until an equilibrium is reached.

8.1 Computational Model and Performance

Here we examine the structural behaviour of a three-blade tidal turbine runner on the fluid loads. Figure 14 shows the mesh around one blade. Since we assume a constant flow distribution at the inlet, the problem is modelled as a periodic section of the whole runner. The open boundaries at the outlet and in radial direction have to be placed sufficiently far away from the runner to avoid influences on the flow solution around the runner.

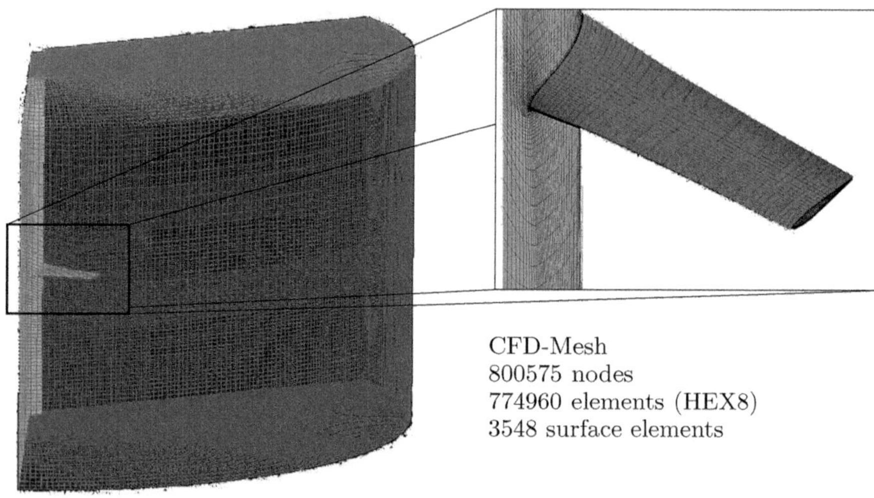

Fig. 14 Coarse periodic CFD mesh (800575 nodes) and detail view of blade surface mesh

Pressure loads to and deflections from the structure are exchanged at the blade surface. First, a CFD analysis without structural interaction is performed on the coarse (0.8 million nodes) and a refined mesh (2.57 million nodes). Table 1 shows a comparison of the performance numbers on a NEC SX-8 vector node (8 CPUs). MPI is applied for parallelisation. Due to the better, but not yet optimal, exploitation of the vector pipes, the refined mesh shows a slightly better performance than the coarse one. All timing values are relative in % to the user-time. Since the code uses unstructured meshes the Bank conflict ratio of about 7% is acceptable. Also the overall code performance of almost 30% peak performance (16 GFlops $\times$ 8 CPUs) including all I/O and the initialisation part is satisfying.

8.2 Fluid-Structure Interaction Results

First, a one-way coupled analysis is conducted, i.e. solving the flow equations and then solving the structure separately with the pressure loads obtained from the fluid

Table 1 Overall average performance data for tidal turbine runner. All times in % relative to user-time

			Vectorisation		Cache Miss	
Mesh size	% peak	Bank conflict	Vec.len.	Vec.ratio	Instruction	Operation
800575	28.06	6.55	248.94	99.44	1.42	1.29
2574740	29.77	7.08	251.55	99.46	1.71	0.87

solution. The results are used to verify the two-way coupling. Here a quasi-static equilibrium between the flow solution with constant inflow boundary conditions and the structural deflection is sought. For this analysis, the sequential explicit staggered scheme, see Fig. 4, left, is applied. The blade material is steel with a Young's modulus of 2.1E+11 N/m^2, Poisson ratio of 0.3 and a density of 7850 kg/m^3.

Figure 15 shows the results of both analyses. In the coupled computation case the pressure distribution (on deflected surface) in the equilibrium state is almost identical to the CFD-only (one-way) solution. In reality the deflection of the blade is rather small; here it is scaled by a factor of 10. This is the reason why the pressure field does not change in the two-way coupled analysis case. In further simulations unsteady boundary conditions will be applied to study the dynamic behaviour of the blade.

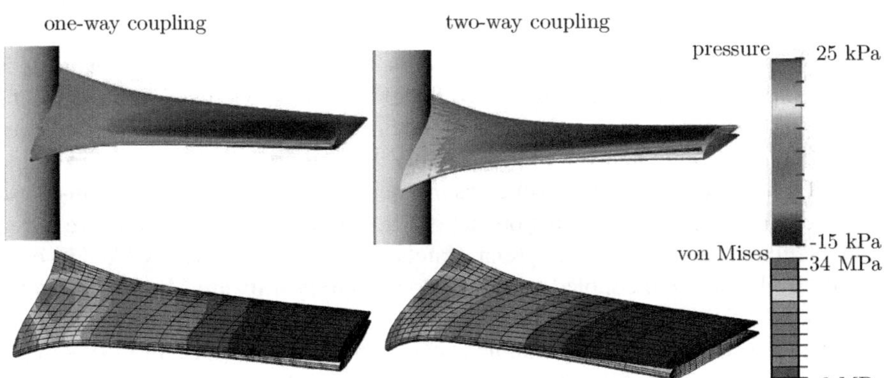

Fig. 15 Pressure field and v. Mises stresses for a one-way coupled solution and a quasi-static two-way coupled solution

Regarding computational time, the coupled FSI analysis takes twice the time of the CFD simulation. Looking at the time needed by the mesh update algorithm this might be surprising. The profiling shows an additional effort of 5.5% of the total simulation time for the new routines. About 3.5% is needed for the initial node search, 2% for the mesh-update after each data transfer. But, there are about 50% MPI_WAIT on the four processors that do not participate in the coupling. On the other four processors that do exchange data, there is almost no waiting time issued. This shows the drawback of the sequentially coupled scheme. The time needed for one step is the same for both codes, but FENFLOSS has to wait until ABAQUS is done with its time-step. To reduce the wasted MPI_WAIT time either the structural code has to run even faster or the coupling scheme has to be changed. Applying the parallel coupling scheme, the MPI_WAIT-time is reduced to 50% in comparison to the sequential scheme. The total ratio is about 33% of the whole computation. This is due to the fact that there is still some time needed for the data exchange and interpolation, and, furthermore, the codes still sometimes have to wait. But the improvement is clear. Obviously, for this kind of simulation the parallel scheme is the best regarding the computation time. It is furthermore desirable to run both codes on platforms that guarantee almost identical step times between the data transfer.

9 Summary

In this article the implementation of a partitioned fluid-structure interaction scheme on different platforms is discussed. The CFD part runs on a SX-8 vector machine, the structural part on a regular cluster. First, the scheme is evaluated with a benchmark example. Second, an adaptive turbulence model is evaluated to estimate the mesh quality, that is necessary to resolve turbulent structures, leading to structural vibrations, accurately. Last, the flexible blades of a tidal current turbine with the surrounding fluid and turbulent flow conditions are analysed.

Due to the increased number of necessary iterations unsteady simulations require more computational power, which is available on vector machines. In case of unsteady fluid-structure interactions in hydraulic machinery, implicit partitioned coupling schemes require an even higher number of solution steps. Here, we examine the performance of the coupling schemes in a quasi-static FSI analysis. The in-house CFD-code FENFLOSS is running on an SX-8 vector node that provides the required computational power, the commercial structural mechanics solver ABAQUS on a cluster. Both codes are coupled with the code coupling software MpCCI that is installed on the SX-8 front end.

After the verification of the sustained vector performance of FENFLOSS a single CFD analysis is conducted. The pressure loads are used to obtain static deflections of a tidal current turbine blade as reference for a quasi-static coupled analysis. For a sequential fluid-structure interaction scheme the simulations show a high ratio of waiting time for the CFD-processes that do not participate in the coupling. A parallel scheme reduces the idle time by 50%.

Future work will be, first, about the improvement of the performance of coupled simulations. Second, dynamic analyses with unsteady boundary conditions will be made to examine the vibrational behaviour of the tidal turbine blades. Furthermore, unsteady vortices and their influence on the structure have to be examined.

References

1. Borowski, S., Tiyyagura, S.R., Küster, U.: Matrix assembly without coloring on vector machines. In: Proc. of the Int. Conference of Numerical Analysis and Applied Mathematics (ICNAAM 2006). Crete, Greece (2006)
2. Cebeci, T.: Turbulence models and their applications. Horizons Publishing, Long Beach, CA (2004)
3. Chen, Y.S., Kim, S.W.: Computation of turbulent flows using an extended $k-\varepsilon$ closure model. NASA CR-179204
4. Donea, J., Huerta, A.: Finite element methods for flow problems. Wiley, Chichester (2003)
5. Farhat, C., Lesoinne, M., le Tallec, P.: Load and motion transfer algorithms for fluid-structure interaction problems with non-matching interfaces. Computer Methods in Applied Mechanics and Engineering **157**, 95–114 (1998)
6. Ferziger, J.H., Perić, M.: Computational methods for fluid dynamics, 3 edn. Springer (2002)
7. Förster, C.: Robust methods for fluid-structure interaction with stabilised finite elements. Ph.D. thesis, Institut für Baustatik, Universität Stuttgart (2007)
8. Gongwer Azusa, C.A.: A study of vanes singing in water. Transactions of the ASME **74** (1952)
9. Gresho, P.M., Sani, R.L.: Incompressible flow and the finite element method, vol. I. John Wiley & Sons (1999)
10. Helmrich, T.: Simulation instationärer Wirbelstrukturen in hydraulischen Maschinen. Ph.D. thesis, IHS, Universität Stuttgart (2007)
11. Hughes, T.J.R., Liu, W.K., Zimmermann, T.K.: Lagrangian-Eulerian finite element formulation for viscous flows. Comp. Methods in Applied Mech. and Eng. **29**, 329–349 (1981)
12. Hübner, B., Seidel, U.: Partitioned solution to strongly coupled hydroelastic systems arising in hydro turbine design. In: Proceedings of the 2nd International IAHR Meeting of the Work Group on Cavitation and Dynamic Problems in Hydraulic Machinery and Systems. Timisoara, Romania (2007)
13. Kjellgren, P., Hyvärinen, J.: An arbitrary Langrangian-Eulerian finite element method. Computational Mechanics **21**, 81–90 (1998)
14. Launder, B.E., Spalding, D.B.: The numerical computation of turbulent flows. Comp. Methods in Applied Mech. Eng. **3** (1974)
15. Lippold, F.: Fluid-structure interaction in an axial fan. HPC-Europa report (2006)
16. Maihoefer, M.: Effiziente Verfahren zur Berechnung dreidimensionaler Stroemungen mit nichtpassenden Gittern. Ph.D. thesis, University of Stuttgart (2002)
17. Menter, F.R.: Two-equation eddy-viscosity turbulence models for engineering applications. AIAA Journal **32**, 1598–1605 (1994)
18. Mok, D.P.: Partitionierte Lösungsansätze in der Strukturdynamik und der Fluid-Struktur-Interaktion. Ph.D. thesis, Institut für Baustatik, Universität Stuttgart (2001)
19. Ruprecht, A.: Finite Elemente zur Berechnung dreidimensionaler turbulenter Stroemungen in komplexen Geometrien. Ph.D. thesis, University of Stuttgart (1989)
20. Schlichting, H., Gersten, K.: Boundary layer theory, 8 edn. Springer (2000)
21. Turek, S., Hron, J.: Proposal for numerical benchmarking of fluid-structure interaction between an elastic object and laminar incompressible flow. In: H-J. Bungartz, M. Schäfer (eds.) Fluid-Structure Interaction — Modelling, Simulation, Optimization, *Lecture Notes in Computational Science and Engineering*, vol. 53, pp.371–385. Springer, Berlin Heidelberg (2006)
22. van der Vorst, H.A.: BI-CGSTAB: A fast and smoothly converging variant of BI-CG for the solution of nonsymmetric linear systems. SIAM Journal of Scientific Stat. Computing **13**, 631–644 (1992)
23. van der Vorst, H.A.: Recent developments in hybrid CG methods. Proc. High Performance Computing & Networking, München (1994)
24. Willems, W.: Numerische Simulation turbulenter Scheströmungen mit einem Zwei-Skalen Turbulenzmodell. Ph.D. thesis, RWTH Aachen, Shaker Verlag, Aachen (1997)
25. Zienkiewicz, O.C., Taylor, R.L.: The finite element method, vol. I. McGraw-Hill (1989)

Heterogeneous Parallel Aero-Acoustics Using PACX-MPI

Harald Klimach and Sabine Roller and Claus-Dieter Munz

Abstract Recent simulations are incorporating more and more aspects of the real world. Each aspect generally needs its own numerical scheme in the simulation. Thus very different numerical requirements for efficient computing are combined within a single simulation. This work looks specifically at the fully integrated simulation of fluid dynamic and aero-acoustic effects. On the architectural side of modern computing hardware there is a diversification going on, a wide range of concepts is considered to gain ever more performance. The two trends result in the need to map the computational requirements of the applications to the available heterogeneous resources efficiently. This work uses PACX-MPI to distribute the distinct parts of the simulation on to clusters with different properties. It is shown how the heterogeneity of both, the computation and the computing resources can be beneficial.

1 The Aero-Acoustic Application

This work considers the application KOP3D [6], developed at the Institute for Aero- and Gasdynamics at the University of Stuttgart. It consists of several parts, which are tightly integrated and allow the coupling of very different computational domains. There is a solver for unstructured grids, a solver for structured grids and the coupling code responsible for interpolation and communication between all domains. Each domain may use a different numerical scheme and can act on different physical equations than its neighbors. This made its usage especially well suited for coupled fluid dynamics and aero-acoustic simulations, where the impact of each part on the other is not neglected but fully taken into account. In the simulation of noise gener-

Harald Klimach · Sabine Roller
HLRS, Nobelstrasse 19, 70569 Stuttgart, e-mail: {klimach, roller}@hlrs.de

Claus-Dieter Munz
IAG, Pfaffenwaldring 21, 70569 Stuttgart, e-mail: munz@iag.uni-stuttgart.de

ated by the flow of a fluid around an obstacle, there are two clearly distinct physical regions. Firstly, there is the flow field which needs to be accurately simulated near the obstacle. Secondly, there is a far field area where the propagation of sound has to be computed. KOP3D enables the coupling of these two regions with appropriate methods on each side. It allows to compute the solution of the Navier-Stokes equations on a high resolved unstructured grid with high order schemes. At the same time the propagation of soundwaves farther away from the body, computed on a coarse structured grid, is coupled with that region.

1.1 Structure of the Application

The KOP3D application is organized in independent, but tightly coupled, parts. The two solvers involved can each be build and run independently of each other. As a stand alone program, they can be used to solve single domains with homogeneous properties like standard solvers. However, they share a large common programming infrastructure and, when compiled together with the coupling part into a single program, are highly integrated. There are a variety of different numerical schemes available in each solver. On the unstructured grid Discontinous Galerkin [2] or Finite Volume methods are usually used. On the structured grid, there are Finite Differences and Finite Volume methods available. All schemes can be run with arbitrary order. The time integration uses the Arbitrary High Order using Derivatives (ADER) scheme. Timestepping of the compressible flow solvers are explicit.

The coupling part acts as the main program, which calls the solvers and controls the complete setup. There may be any number of different domains coupled together, each with its appropriate solver. Coupling of the different domains is realized by taking care of the boundary conditions between neighboring domains. The taken approach of discrete coupling between different domains allows different discretizations, numerical schemes and even equations in adjacent domains and retains the high order of involved schemes across domain boundaries. Moreover, there is a subcycling strategy to allow different timesteps between neighbors. This results in domains with potentially rather small interactions between each other. Nevertheless, there is a tight integration into a single program.

Boundaries are realized by ghost cells outside the computational area itself. On physical boundaries the ghost cells are prescribed in way which satisfies the boundary conditions. At interfaces between adjacent domains those ghostcells are used to capture the solution from the corresponding neighbor.

1.2 Coupling Scheme

As the solution in different domains is exchanged by ghost cells, these cells have to be constructed in a way, which allows the interpolation between the domains. The ghostcells are conforming to the grid layout of the domain they are belonging to and fit naturally into its spatial discretization. Therefore, no special care needs to be taken on this receiving side of the coupling. On the other side, since there are no restrictions on the layout of the domains in relation to each other, there may be ghost cells which are spread across more than one neighbor. The exchange of the solution has to be ensured for any geometric layout covered by the ghost cells. To achieve this, the coupling scheme uses discrete point values from the neighbor domains. This allows a unique identification of the neighbor domain, where the value has to come from, and thereby allows the separation of data collection and actually constructing the cells on top of the given points. This is possible, as the used numerical schemes can be reduced to data at discrete points to describe the state in each cell. The burden of building the ghost cell is by this in the responsibility of the domain needing them. Neighboring domains just need to provide the values at requested points.

As for very different cell sizes in adjacent domains different timesteps are implied by the stability criterion, there is a mechanism needed to allow these. In the application of this work a subcycling scheme is used, where all domains have timesteps which are a multiple of the smallest timestep in any domain. In order to allow the domains with smaller timesteps to progress, a Taylor series in time is used to estimate the solution at each timelevel and each point based on the last common timestep with the neighbor. The Taylor series is obtained by translating the space derivatives to time derivatives and applying the Cauchy-Kowalevsky procedure [4] as in the ADER time integration scheme.

1.3 Parallelization

The parallelization of the coupled application is solely based on MPI. There is a natural parallelization by the distribution on several domains, which all can be computed in parallel due to the explicit timestepping. This domain level parallelism is exploited by the MPI parallel coupling scheme. Another level of parallelism arises from the fact, that each solver itself is parallelized using MPI and thus allowing its computation to be spread across several processes. In coupled simulations, the coupling part takes care of the distribution of domains onto different processes. It assigns an appropriate number of processes to each domain and each solver just acts like it would have, when started with the given number of processes on its own. The distribution of processes can be arbitrarily chosen and one process may be responsible for more than one domain or just compute a fraction of a large domain.

The data separation between neighboring domains, enabled by the two steps in the construction of ghost cells at interfaces, is an important feature for the parallel version of the program. Only lists of point coordinates have to be sent to neighbors and lists of values at the corresponding points are to be received, because of this feature. As the domains may have different timesteps using the subcycling mechanism and only on matching timesteps data has to be exchanged, the interaction link between domains with heavily differing grid resolutions is usually rather weak. Due to this fact, communication between most of the neighbors is greatly reduced, compared to the exchange of partitions within each domain, which has to be done every timestep.

2 PACX-MPI

PACX-MPI [3] is an MPI layer developed at HLRS which enables an MPI parallel application to be spread across several clusters [1]. For the application this distribution is hidden by PACX-MPI, which acts as an intermediate layer between the application and the actual MPI libraries on each involved cluster.

2.1 Communication Layout

PACX-MPI takes care of the communication between different clusters by using one sending and one receiving daemon on each of them. Necessary communication between processes residing on different machines is intercepted by PACX-MPI and tunneled through the daemons to the corresponding machine. Communication of processes on the same cluster is handed over to the local MPI implementation. With this approach the optimal usage of the networking infrastructure on each side is available to the application and PACX-MPI is free to use any different appropriate protocol to establish the inter cluster link. This layout of the process distribution within PACX-MPI with the communication daemons and the mapping of local MPI ranks to global MPI ranks is shown in Fig. 1.

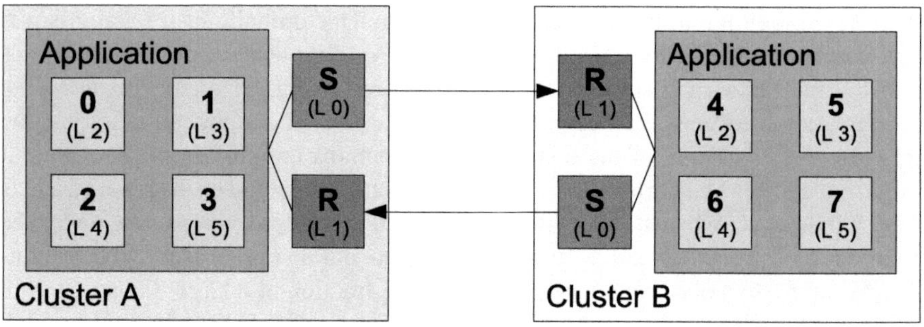

Fig. 1 Layout of the PACX-MPI communication across distinct clusters

2.2 Heterogeneous Environment

The PACX-MPI library may be used to link smaller similar clusters together to gain more computational power for a single simulation, than on each of them may be available. But PACX-MPI is also capable to couple very different kind of machines together. For example it transparently takes care of necessary data conversions from little endian to big endian. And as PACX-MPI is providing the usual MPI to the application transparently it provides a homogeneous setting on top the heterogeneous hardware. This yields the advantage to have the possibility to distribute any MPI application across a heterogeneous environment, however it should be kept in mind, that, in order to use the hardware efficiently, the application should be aware of the weak links between different clusters nevertheless.

2.3 Starting an Application Using PACX-MPI

There are two possibilities to start an application using PACX-MPI. The complete setup may be defined by hand in configuration files on each site. This approach is a very static and strict one, but the complete setup can be defined in every detail. The more dynamic approach is to use the provided startup server. The startup server has to be started before the application instances are executed and listens on a specified port. When the application processes are started, they send their setup informations to the startup server which collects them and sends all informations back to each process after the last instance is connected. The startup server is then terminated. This approach is much more flexible as the application just has to have the information where the server is listening and the server just needs to be informed of the number of intended application instances, which is a command line argument.

3 Heterogeneous Computations

The heterogeneous supercomputing environment used in this work is the NEC SX-8 vector machine at HLRS and its IA-64 cluster front end. A 3D test case where a sinusoidal wave is scattered at a perfectly reflecting sphere of one meter radius is used to analyze the behaviour of KOP3D in this setting. The exact solution for this problem is known and was proposed as an aero-acoustic benchmark by Morris for the *Second CAA Workshop on Benchmark Problems* [5]. The space around the sphere is discretized with a unstructured mesh using Discontinous Galerkin methods and is embedded into a structured mesh where a Finite Differences scheme is deployed.

The unstructured mesh consists of 9874 elements and the structured mesh of around 42 million. In total a volume of $102.2 \times 57 \times 57$ meters is covered in the simulation. This configuration of domains is shown in Fig. 2 with the pressure distribution after two seconds of simulation time as coloring.

The complete simulation was done on each architecture for comparison and then each domain was run on the appropriate machine. The distributed run, of course, has to be done simultaneously on both machines. Ideally this would be ensured by the scheduler, however this feature is not yet available in the current infrastructure of this setup. The simultaneous execution on both parts therefore had to be ensured manually. This was done by starting the computation on the cluster interactively and submitting a job for the part on the vector machine to the scheduler. For the initialization of the PACX-Setup, a startup server running on the cluster was used.

The runtimes on each architecture and in the coupled run are shown in Table 1. By looking at the total elapse time on each architecture, we see an advantage of almost a factor of three, if the computation is done on the vector architecture instead of the scalar cluster. But the table also highlights that the unstructured part of the simulation is burning most of the time on the vector machine, and takes more than twice as long as on the scalar architecture. As the vector processor are much more expensive, the computation of the unstructured grid is a waste of resources. The 42 million cells in the structured far field of the simulation are a heavy workload for the scalar architecture and need a lot of memory. Obviously we would like to put this

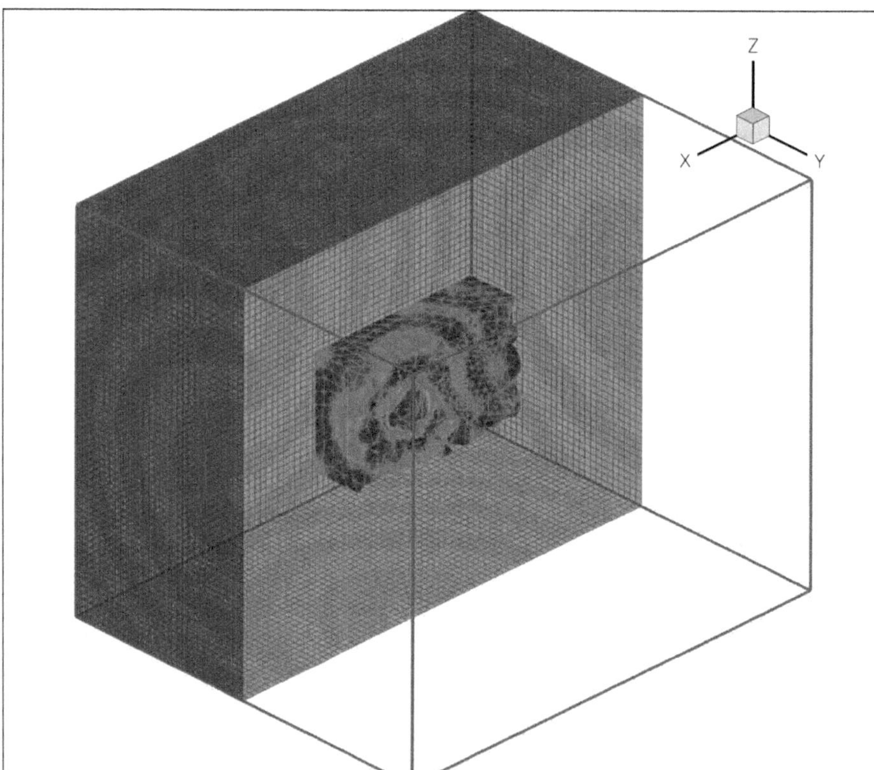

Fig. 2 Pressure distribution in the two involved domains around the perfectly reflecting sphere

Table 1 Runtimes on different architectures

	Itanium II	NEC-SX8	Both coupled
Elapsed time	27925 s	10966 s	3207 s
Unstructured	2994 s	7746 s	3019 s
Structured	23887 s	2871 s	2869 s
Coupling	1012 s	321 s	554 s
Waiting	0 s	0 s	164 s

heavy load on the vector machine in order to get the result for this 2 seconds of simulation time in a reasonable time. PACX-MPI gives us the power to do so and take advantage of the strengths of both architectures. By distributing the coupled simulation with its different requirements on the heterogeneous hardware environment, time to solution is drastically reduced.

Each part of the simulation is executed on the platform which offers the highest performance for it. Thus a short elapse time is reached despite the introduced MPI waiting times between both domains. The waiting times are a direct result of the imbalance in the runtimes of the solver on each side. A perfect balance can not be gained, because of the two types of involved architectures and the number of CPUs which necessarily have to be in whole numbers, but this problem may be neglected for larger numbers of processors. However, this imbalance can be kept small for larger numbers of CPUs.

4 Conclusion

As shown with this coupled fluid dynamic / aero-acoustic application, it is possible to gain advantages on different computational architectures by using PACX-MPI without changing the application itself. Deploying PACX-MPI to distribute the computations in a heterogeneous infrastructure is easily done by relinking the application and using the startup server during the initialization of the MPI run. The deficits of this coupling over weaker links in the communication, need for data conversions, and newly introduced imbalances are outweighed by the greatly enhanced overall efficiency. By using PACX-MPI and architectures with different features, the time needed to solve a problem can be greatly improved without the need to optimize each part of the simulation for any architecture. When the co-scheduling between distinct clusters becomes available, this heterogeneous approach can be an option for immediate speed-up of coupled applications with various numerical requirements.

References

1. Gabriel, E., Resch, M., Beisel, T., Keller, R.: Distributed computing in a heterogeneous computing environment. In: Proceedings of the 5th European PVM/MPI Users' Group Meeting on Recent Advances in Parallel Virtual Machine and Message Passing Interface, *Lecture Notes In Computer Science*, vol. 1497, pp. 180–187 (1998)
2. Hesthaven, J.S., Warburton, T.: Nodal Discontinuous Galerkin Methods: Algorithms, Analysis and Applications. Springer (2007)
3. Keller, R., Müller, M.: PACX-MPI project homepage. http://www.hlrs.de/organization/amt/projects/pacx-mpi
4. v. Kowalevsky, S.: Zur Theorie der partiellen Differentialgleichungen. Ph.D. thesis, Philosophische Fakultät zu Göttingen (1874)
5. Morris, P.: Scattering of sound by a sphere: Category 1, problems 3 and 4. In: Proceedings of Second Computational Aeroacoustics (CAA) Workshop on Benchmark Problems (1997). NASA/CP-1997-3352
6. Utzmann, J., Schwartzkopf, T., Dumbser, M., C.-D. Munz: Heterogeneous domain decomposition for computational aeroacoustics. AIAA Journal **44**, 2231–2250 (2006)

Meandering of Wing-Tip Vortices Interacting with a Cold Jet in the Extended Wake

Frank T. Zurheide, Matthias Meinke, and Wolfgang Schröder

Abstract The spatial development of a vortex wake behind a half wing is simulated up to the extended near field. Results from wind tunnel measurements are used as inflow boundary conditions for an LES of the wake region. An aircraft engine is modeled in the experimental setup, where a cold jet is driven by pressurized air. The engine is mounted in two different positions under the wing model to investigate the influence of the location of the engine jet on the vortex wake. The numerical simulations of the wake are able to predict trajectories and instabilities of the vortex core. A position of the engine towards the root of the wing creates a smaller deflection of the vortex and excites fewer wave modes of the vortex.

1 Introduction

The limitations of airport capacity due to wake vortex fields is of growing importance due to the expected increase in passenger numbers and aircraft traffic. Heavy transport aircraft generate vortex wakes which endanger the following aircraft, because they cannot be precisely detected. The decay of vortex wakes by diffusion and dissipation mechanisms is weak, such that the frequency of starting and landing aircraft is limited by waiting times required for a sufficient decay of the vortex wake. Instabilities of wake vortex systems, however, can lead to a rapid decay of the vortex strength. Scientific research focuses on the evolution of instabilities inherent to or artificially introduced into the wake structure [44]. The aim is the excitation of mechanisms which lead to an amplification of different types of instabilities and therefore reduce the time required for the onset of rapid vortex decay.

In vortex systems several three-dimensional instabilities are known. Mutual straining of the vortices leads to an amplification of asymmetric Kelvin waves if

Frank T. Zurheide · Matthias Meinke · Wolfgang Schröder
Institute of Aerodynamics, RWTH Aachen University, Wüllnerstr. 5a, 52062 Aachen, Germany, e-mail: {f.zurheide, m.meinke, office}@aia.rwth-aachen.de

three-dimensional perturbations exist. Due to this amplification, vortex systems are generally unstable. Using a few assumptions concerning vortex core radius and vortex spacing, the vortex system behind wings can be regarded as a set of parallel vortex filaments, the perturbations are Kelvin waves with a small amplitude. From this model a system of stability equations can be derived. Crow [8] presented the results of this linear system for a single pair of counter rotating vortex filaments. Crouch [7] and Fabre et al. [10, 11] performed the stability analysis for multiple vortex pairs. The wave length of the Crow instability is $\lambda/b \approx 7.85$ [8] and following Holzäpfel et al. [18], Chap. 8.6.

Stability analyses of a vortex pair using the vortex filament method result in basic findings for short-wave elliptical instabilities, that have a higher growth rate of instabilities than the long-wave instability were presented by Waleffe [47]. The existence of this instability has been shown in experiments for counter-rotating vortex pairs by Leweke and Williamson [30] and co-rotating vortex pairs by Meunier et al. [36, 37, 35]. Numerical simulations of this instability were performed by Laporte and Corjon [27], Laporte and Leweke [28] as well as Le Dizès and Laporte [29]. Sipp [43] studied the nonlinear dynamics of the instability. Viscous effects change the behavior of the vortex system. The sinusoidal instability saturates at a low level and does not lead to a breakdown of the system. Consequently the slower growing long-wave instability (Crow instability) are of prime interest.

One possible strategy to influence the vortex wake is to use the engine jet. There are, however, only few articles that analyze the interaction of an engine jet and a trailing vortex in the near field of the wake. The growth of instabilities in a jet are well understood. Loiseleux et al. [32] analyzed the effects of swirl on jets and wakes and Lu and Lele [33] the inviscid instabilities for swirling mixing layers.

Miake-Lye [38] identified two distinct phases of the jet/vortex interaction. In the first phase, the jet regime, the engine jets mix with the ambient air and the vortices roll up to a vortex pair. In the second phase, the interaction regime, entrainment of jets into the vortex flow dominates. In contrast to that model, experiments from Brunet et al. [6] showed that at a distance of only one semispan downstream of the wing the jet was already entrained to a certain extent.

The influence of the jet on the tip vortex was analyzed with temporal simulations by Labbé et al. [26], Paoli et al. [40], Ferreira Gago et al. [14] and Holzäpfel et al. [20]. Unlike these temporal investigations, Stumpf [45, 46] and Fares and Schröder [12] simulated a spatial developing wake/jet interaction in the near field based on the Reynolds averaged Navier-Stokes equations.

Vortex meandering, also known as vortex wandering, was introduced by Baker et al. [2] in 1974. Devenport [9] developed strategies to filter out the unsteadiness and to correct the time-averaged vortex position. Jacquin et al. [23] explains vortex meandering by the amplification of bending waves through external freestream fluctuations and linear co-operative instabilities. Beninati and Marshall [3] found no evidence for vortex meandering, but a bending wave resulting from the interaction with vortex turbulence.

In this paper we will present a large-eddy simulation (LES) of a spatially developing wake-jet interaction in the near field and extended near field. Measurements from the wake are used as inflow data for the simulations. The flow fields for two different positions of an engine jet are analyzed and compared.

This paper is organized as follows. The governing equations and the numerical procedure of the LES method are described in Sect. 2. The inflow data for the simulations are presented in Sect. 3. The results for the wake in the extended near field is given for two different positions of the engine and is discussed in detail in Sect. 4. The computational resources used for the simulations are described in Sect. 5. Finally, major the findings of the present study are summarized in Sect. 6.

2 Governing Equations

Since the continuum assumptions hold for the flows regarded here, the Navier-Stokes equations are an appropriate mathematical model for their simulation. Written in tensor notation and in terms of dimensionless conservative variables for a Cartesian coordinate system, they read

$$\frac{\partial \mathbf{Q}}{\partial t} + (\mathbf{F}^C_\beta - \mathbf{F}^D_\beta)_{,\beta} = 0\,, \qquad \mathbf{Q} = [\rho, \rho u_\alpha, \rho E]^T\,, \tag{1}$$

where $\mathbf{Q}$ is the vector of the conservative variables, and $\mathbf{F}^C_\beta$ represents the vector of the convective and $\mathbf{F}^D_\beta$ of the diffusive fluxes

$$\mathbf{F}^C_\beta - \mathbf{F}^D_\beta = \begin{pmatrix} \rho u_\beta \\ \rho u_\alpha u_\beta + p\delta_{\alpha\beta} \\ u_\beta(\rho E + p) \end{pmatrix} - \frac{1}{Re} \begin{pmatrix} 0 \\ \sigma_{\alpha\beta} \\ u_\alpha \sigma_{\alpha\beta} + q_\beta \end{pmatrix}.$$

The stress tensor $\sigma_{\alpha\beta}$ is written as a function of the strain rate tensor $\mathbf{S}_{\alpha\beta}$, i.e.

$$\sigma_{\alpha\beta} = -2\mu(\mathbf{S}_{\alpha\beta} - \tfrac{1}{3}\mathbf{S}_{\gamma\gamma}\delta_{\alpha\beta})$$

with

$$\mathbf{S}_{\alpha\beta} = \tfrac{1}{2}(u_{\alpha,\beta} + u_{\beta,\alpha})\,.$$

The dynamic viscosity μ, is assumed here to be only a function of the temperature. Fourier's law of heat conduction is used to compute the heat flux q_β as

$$q_\beta = -\frac{k}{Pr(\gamma - 1)} T_{,\beta}\,,$$

where Pr is the Prandtl number. The system is closed using the equation of state for ideal gases

$$p = \frac{1}{\gamma}\rho T\,, \quad p = (\gamma - 1)\left(\rho e - \frac{\rho}{2} u_\beta u_\beta\right)\,,$$

where γ is the ratio of the specific heat capacities and T the temperature.

For the LES, an implicit grid filter of width Δ is assumed to be applied to Eq. (1). A sub-grid scale model and the truncation error are coupled to the spatial steps of the computational grid in case of a finite volume scheme. Both terms lead to a damping of the smallest resolved scales and can be shown to be of the same order of magnitude for second-order accurate schemes, at least in the time averaged sense [17]. For the solution scheme presented in the next section, the influence of the sub-grid scale model has a negligible influence on accuracy of the turbulence statistics [34]. Therefore, no sub-grid scale model was used for the simulations presented here.

3 Boundary Conditions

The wake is measured behind a half model of a wing below which an engine can be mounted at different positions. The results of the particle image velocimetry (PIV) measurements are used as inflow boundary conditions for the spatial computation of the wake in the near field and in the extended near field.

3.1 Experimental set-up

The experiments were carried out in the low-speed wind tunnel of the Institute of Aerodynamics, RWTH Aachen University [21, 22]. The flow speed is set to $u_\infty = 27$ m/s which results in a Reynolds number of $Re_c = 2.8 \times 10^5$ based on the mean chord length c_m. The wing has a BAC 3-11/RES/30/21 profile and a three dimensional wing-tip geometry. The semi span is $s = 607$ mm, and the mean chord length is $c_m = 150$ mm. The angle of attack is $\alpha = 8^\circ$.

An engine model can be mounted at two positions below the wing. The engine model works as an air jet apparatus and is operated with pressurized air. The jet speed is approximately $u_{engine}/u_\infty = 1.74$.

The stereoscopic particle-image velocimetry (PIV) method is used for the measurement of the wake velocity field. The three velocity components of the wing and engine wake are captured with stereo- or 3C-PIV one chord length c_m behind the wing (see Fig. 1). The measurements were carried out for two positions of the jet, position “A” towards the wing tip and position “C” towards the root of the wing. The results of the measurements are recorded on a Cartesian mesh with 480×156 points, where the dimensions of the measurement plane are 453 mm $\times$ 215 mm.

In Fig. 2 the Mach number distributions of the measurements are shown for the different engine positions. In Fig. 2a the jet can be recognized by the high velocity i.e. dark region centered at $y/b \approx 0.2, z/b \approx -0.01$. The shear layer of the wing can be identified by the lower velocity, i.e. the brighter thin region that starts at the left border at $z/b \approx 0.02$ and is entrained at the right by the tip vortex. The center of the vortex is located at $y/b = 0.35$, $z/b = 0.022$. For both engine positions, the same

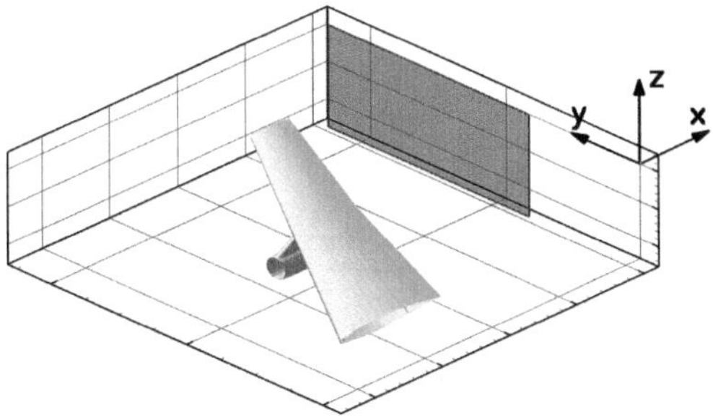

Fig. 1 Schematic of measurement set-up. BAC 3-11 wing with the engine mounted in position "A"/"C". The dark plane behind the wing indicates the location of the PIV measurements

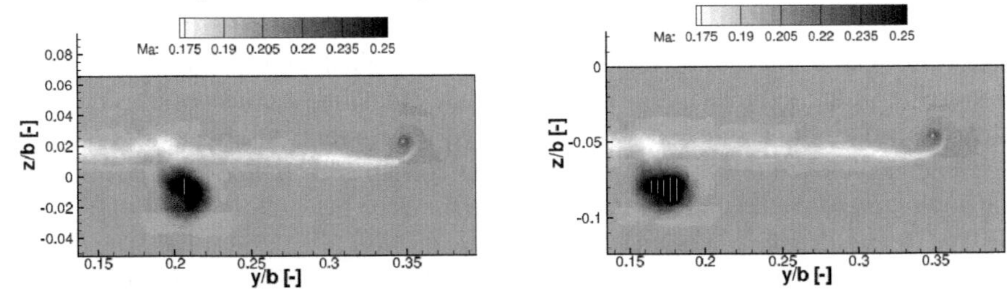

a Engine position "A" **b** Engine position "C"

Fig. 2 Mach number distribution for engine positions "A" and "C"

features (jet, shear layer and tip vortex) can be seen, but the jet for engine position "C" is moved towards the root of the wing.

3.2 Inflow boundary conditions

The results of the PIV measurements are used as inflow boundary conditions of the numerical simulation. The extent of the measurement window in spanwise and wing normal direction is not large enough to permit a simulation of the interaction of the engine jet and the tip vortex, so the measurement data is interpolated to a larger area. For the interpolation, a parallel flow with $Ma_\infty = 0.2$ far away from the measurement window is used. The measured data comprise the three velocity components u, v, w and the fluctuating components u', v' and w'. From the velocities the initial pressure distribution p is computed by solving the Poisson equation for the pressure

$$\nabla^2 p = -\frac{\partial u_\alpha}{\partial x_\beta}\frac{\partial u_\beta}{\partial x_\alpha} \tag{2}$$

with the Laplacian operator ∇^2.

From the standard deviation σ of the fluctuating velocity components u', v', w', random fluctuations with the same amplitude are superimposed to the time averaged velocity components $\bar{u}, \bar{v}, \bar{w}$ at the inflow boundary during the simulation.

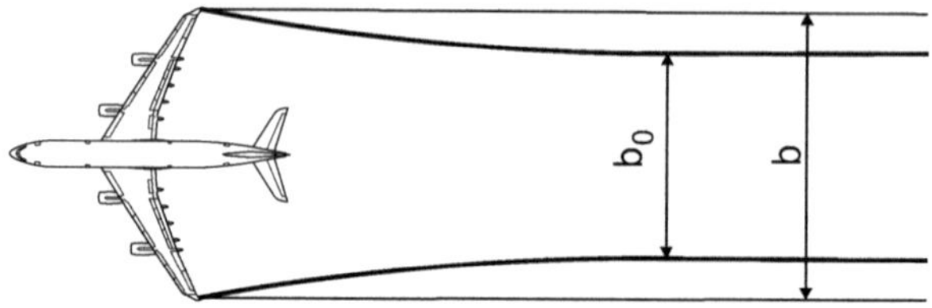

Fig. 3 Schematic wake of an airplane in the extended near field. The wake rolls up to a two-vortex system

4 Results

The wing span $b = 11.56\ c_m$ can be regarded as a typical length scale of the simulation of the wake in the near and extended near-field, since the instabilities in the far field depend on the distance b_0 between the vortex centers. A schematic wake of an airplane is sketched in Fig. 3. The tip vortices roll up and change their position to a separation b_0 in the far-field of the wake [42].

The Reynolds number based on the wing span is $Re_b = 3.2368 \times 10^6$. The domain of integration has a size of $x/b \times y/b \times z/b = 12.48 \times 0.35 \times 0.34$ with the x-axis pointing downstream (see also Fig. 1). The length of the domain is $144c_m = 12.48b$. Within that domain only one half of the wake of the airplane is simulated. The number of mesh points is $1249 \times 225 \times 225$ equally decomposed into 24 blocks leading to 64 million points.

The radius of the wing tip vortex is $r_c = 1.93 \times 10^{-3}b$, which is 44 times smaller than the radius of the vortex used by Holzäpfel et al. [18, 20, 19] or Zurheide and Schröder [49], about one order of magnitude smaller than the radii of the vortices used by Fabre and Jacquin [10], Jacquin et al. [23] and nearly two orders of magnitude smaller than the vortices examined by Dizès and Laporte [29] and by Bristol et al. [5].

While Labbé et al. [26], Paoli et al. [40], and Gago et al. [14] use a temporal approach for the simulation of a jet/wake interaction, we use a spatial approach described by Fares and Schröder [12] and Stumpf [45, 46]. In the temporal approach, the velocity deficit in the vortex core in axial direction is neglected although it is known to have a strong effect on the absolute and convective instability of a vortex [48]. Additionally, the deformation of the vortex is not convected downstream and parts of the wrapped engine jet around the tip vortex have an influence on the flow field upstream, since periodic boundary condition in axial direction are used for the temporal development. On the other hand the spatial approach allows only relatively short extents of the wake to be computed due to the vast number of mesh points for a sufficient resolution of the computational domain.

In Fig. 4, the wake of the simulations at engine position "A" is displayed for two different non-dimensional time levels $t = 29.0$ and 29.5. The wake is shown from a position inside the wake towards the wing. The thick dark tube is a contour of the λ_2-criterion [25] that visualizes the vortex. Streamlines are used to illustrate the jet development. The vortex moves towards the root of the wing, while the jet

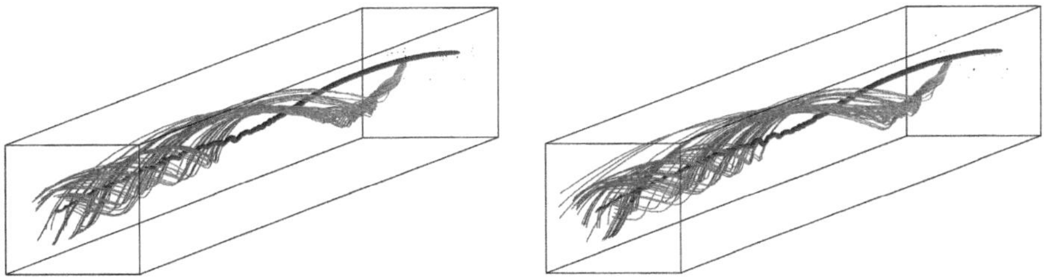

a $t = 29.0$ **b** $t = 29.5$

Fig. 4 Wake of computed measurements with engine position “A”. λ_2-criterion is used to display the vortex (thick dark line) and streamlines to show the curvature and position of the jet. The main flow direction is from top right to bottom left. The symmetry plane is on the left side

is wrapped around the vortex. The vortex reaches the maximal deflection behind the wing at $x/b \approx 4$. Regarding symmetry the deflection is $y/b = 0.8$ for engine position “A” and $y/b = 0.84$ for position “C” shown in Fig. 12. This is close to the theoretical deflection behind a wing with elliptical circulation distribution of $y/b = \pi/4 = 0.785$ [42].

When the jet is wrapped a half rotation around the vortex small perturbations can be identified on the vortex tube (at $x/b = 3$). These perturbations change their shape and grow to a maximum size at $x/b = 5\ldots7$ and are convected downstream while they are decaying. The perturbations have a 3D shape and turn clockwise with the rotation of the vortex.

The change of the shape of the disturbances can be explained by looking at the incompressible form of the azimuthal vorticity equation

$$\frac{D\omega_\theta}{Dt} = \omega_r \frac{\partial v_\theta}{\partial r} + \frac{\omega_\theta}{r}\frac{\partial v_\theta}{\partial \theta} + \omega_x \frac{\partial v_\theta}{\partial x},$$

where $D(\cdot)/Dt$ is the total derivative. Shortly downstream of the wing the radial and azimuthal vorticity components are zero $\omega_r = \omega_\theta = 0$. Axial variations of the azimuthal velocity component are induced when the jet is wrapped around the vortex. This leads to an axial vorticity distribution which is transferred to an azimuthal vorticity $\partial\omega_\theta/\partial t \approx \omega_x \partial v_\theta/\partial x$.

In Fig. 5, the wake of the wing with the engine position “C” is shown. Since the initial distance between the vortex and the jet is larger than for position “A” the interaction between them occurs further downstream. The comparison of the two wakes for engine positions “A” and “C” shows the perturbations at position “A” to have a slightly bigger amplitude and to occur a little further upstream.

In Figs. 6 and 7, the vorticity ω_x for several locations in streamwise direction is displayed. Figure 6a contains the wing tip vortex and the wake. The wake and the jet are wrapped around the vortex for the slices downstream (Figs. 6a–f). From $x/b = 2.0$, one part of the jet can be seen at $y/b = 0.3$, $z/b = -0.07$ for engine position “A” in Fig. 6d. The jet is moved to $y/b = 0.3$, $z/b = 0.016$ at $x/b = 6.0$ (Fig. 6f). For engine position “C” in Fig. 7, the vortex and the wake show a similar behavior

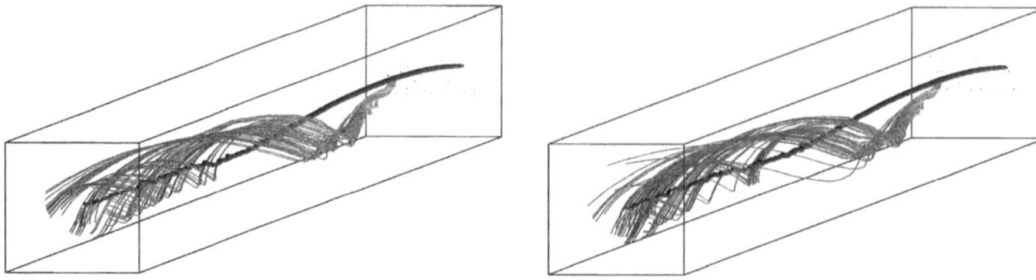

a $t = 29.0$ **b** $t = 29.5$

Fig. 5 Engine position "C". Displayed values, see Fig. 4

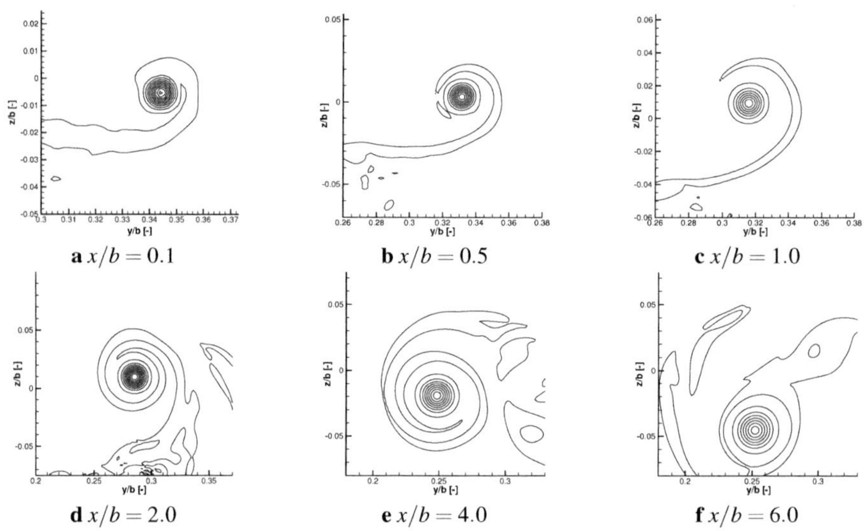

Fig. 6 Vorticity ω_x for engine position "A" at different x positions. 11 Levels for each plot. $x/b = 0.1 : 0.5 < \omega_x < 45$, $x/b = 0.5, 1.0 : 0.5 < \omega_x < 30$, $x/b > 2.0 : 0.2 < \omega_x < 16$

as for position "A". At $x/b = 2.0$ in Fig. 7d, the jet is at $y/b = 0.25$, $z/b = -0.1$, which is outside the picture. Then, at $x/b = 6.0$, the jet is stretched to a line from $y/b = 0.37$, $z/b = -0.01$ to $y/b = 0.32$, $z/b = 0.0$.

To analyze the stability of the jet/vortex interaction the axial Fourier mode k is calculated. To determine the growth of the modes of the instabilities the root of the kinetic energy $E^{1/2} = (1/2(u^2 + v^2 + w^2))^{1/2}$ is used to compute the discrete Fourier coefficients $E_k^{1/2}$ in the yz-plane for each grid point. The coefficients $E_k^{1/2}$ are subsequently averaged over each plane and plotted as a function of the streamwise direction. The discrete Fourier coefficients are calculated using the Fast-Fourier Transformation (FFT).

In Fig. 8, the time averaged Fourier coefficients for the modes $k = 2$ and 4 are shown for the two engine positions. The decay of the noise added at the inflow can be seen in Figs. 8a and 8b for $x/b < 2$. The decay of the kinetic energy of position "A" is stronger than that for position "C". In position "A", the second mode

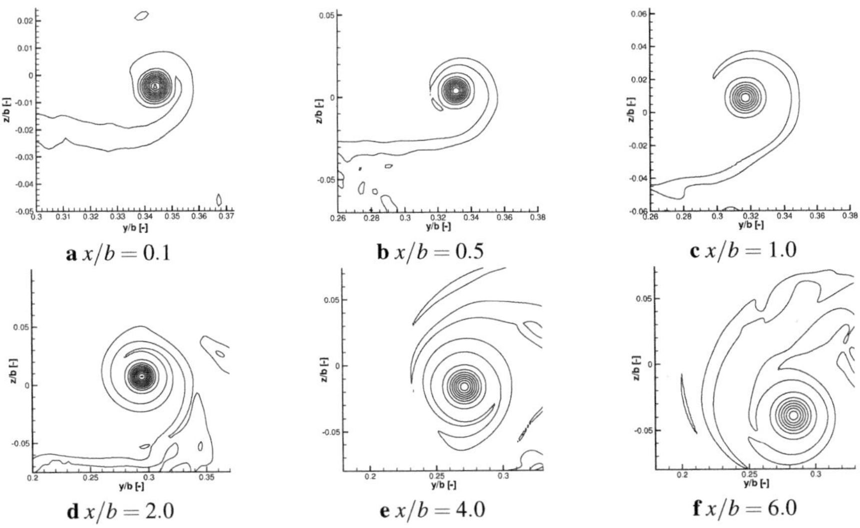

a $x/b = 0.1$ **b** $x/b = 0.5$ **c** $x/b = 1.0$

d $x/b = 2.0$ **e** $x/b = 4.0$ **f** $x/b = 6.0$

Fig. 7 Vorticity ω_x for engine position "C" at different x positions. 11 Levels for each plot. $x/b = 0.1 : 0.5 < \omega_x < 45, x/b = 0.5, 1.0 : 0.5 < \omega_x < 30, x/b > 2.0 : 0.2 < \omega_x < 16$

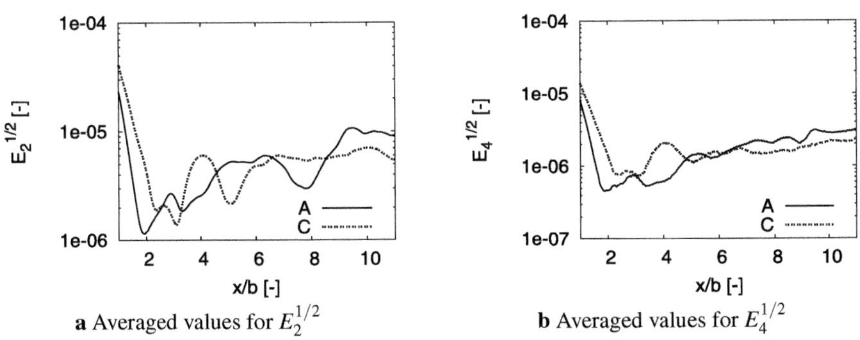

a Averaged values for $E_2^{1/2}$ **b** Averaged values for $E_4^{1/2}$

Fig. 8 Averaged Fourier coefficients $E_2^{1/2}, E_4^{1/2}$ along x-axis. Fourier coefficients are calculated with 64 points

stops to decline and starts to grow at $x/b = 1.93$. This inflection point is at engine position "C" at $x/b = 2.36$ for the first small peak and at $x/b = 3.12$ for the start of the growth with the peak at $x/b = 4.08$. The growth rate of position "C" is larger than for position "A". Further downstream the Fourier coefficients of the energy for position "A" are larger than for position "C".

Compared to the temporal approaches of Paoli et al. [40], Laporte and Corjon [27], Jacquin et al. [23], or Zurheide and Schröder [49], no dominant growth of a mode can be found in the simulated domain $x/b < 12$ which implies that no growth of an short or long-wave instability can be found. Areas with lower Fourier coefficients change with areas of higher values of the Fourier coefficient.

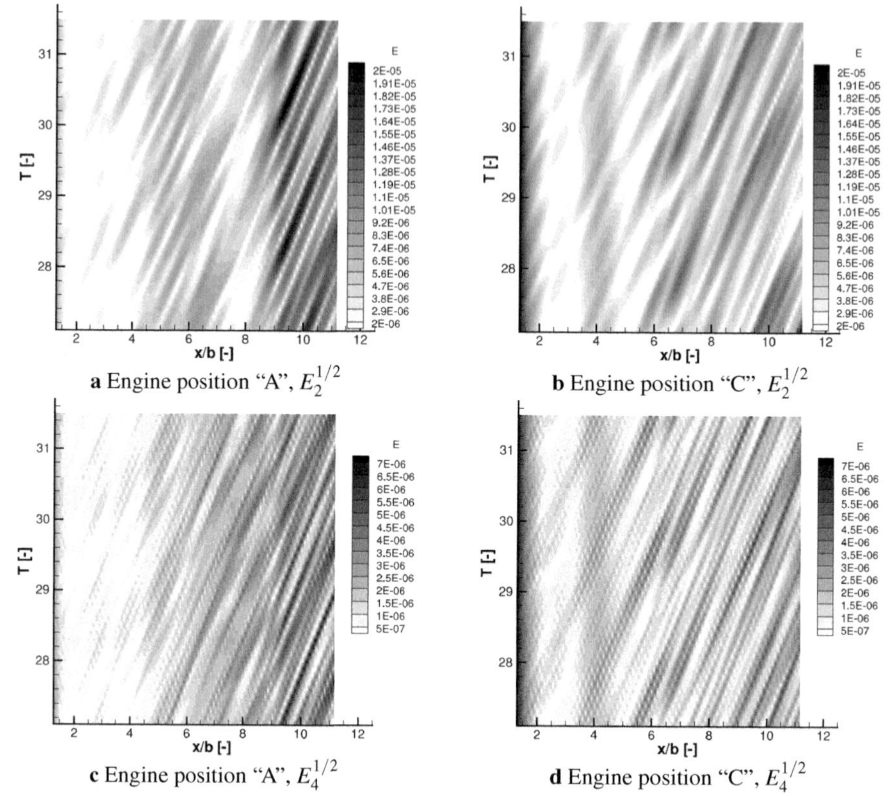

a Engine position “A”, $E_2^{1/2}$

b Engine position “C”, $E_2^{1/2}$

c Engine position “A”, $E_4^{1/2}$

d Engine position “C”, $E_4^{1/2}$

Fig. 9 Fourier coefficients $E_2^{1/2}$ and $E_4^{1/2}$ along x-axis for non-dimensional time levels $t = 27.0 \ldots 31.4$. Fourier coefficients are calculated with 64 points

There are several explanations for this behavior. First, the growth of an elliptical instability requires two vortices that interact but we are computing only one half of the wake with one vortex. Note that a mirror condition at the symmetry plane $y = 0$ would not result in a proper flow condition since the short wave instabilities have an offset of half the wave length [5, 30]. Secondly, the simulated extent in down stream direction is too small. The long wave Crow instability appears several hundred wing spans behind an airplane with large span [16, 41].

For the Fourier modes $k = 2$ and 4, the temporal development is displayed in Fig. 9 for the two engine positions. Unlike the averaged coefficients in Fig. 8, single perturbations of the flow field can be identified.

In Fig. 9a, the Fourier coefficient has at $x/b = 8.0, t = 27.0$ a local maximum. The disturbance is convected downstream and is growing, e.g., at $x/b = 9.0, t = 28.0$ and at $x/b = 11.0, t = 30.0$. At a fixed x position the values show a heterogeneous behavior. The displayed values are striped — high values are followed by lower values. A comparison of the values of engine position “A” (Fig. 9a) and “C” (Fig. 9b) show, that position “A” is more uniform, while the changes between lower and higher val-

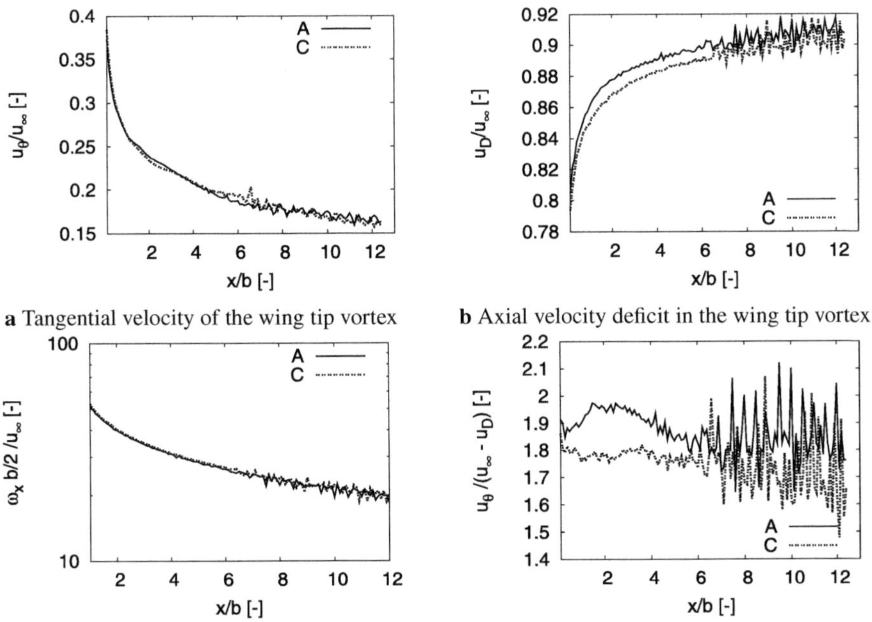

a Tangential velocity of the wing tip vortex

b Axial velocity deficit in the wing tip vortex

c Stream wise vorticity $(\omega_x b/2)/u_\infty$

d Swirl parameter q

Fig. 10 Comparison of values for engine position “A” and “C”

ues at position “C” are more random. Positions where the Fourier coefficients grow can be identified by transitions from light to dark regions, e.g., for case “A” $E_2^{1/2}$ at $x/b = 9.0$ in Fig. 9a or for “C” at $x/b = 4.0$ in Fig. 9b. These peaks can also be found in Fig. 8a. The Fourier coefficients for $E_4^{1/2}$ in Figs. 9c and 9d show a similar characteristic of the convection of flow features, but the growth is lower.

For further investigations of the wing tip vortex, the flow field is analyzed at discrete x coordinates having an increment of $\Delta x/b = 0.1$. The values are taken at a constant dimensionless time $t = 30.0$. The dimensionless tangential velocity u_θ/u_∞ of the wing tip vortex is compared for the two engine positions (Fig. 10a). The difference between the decay of the tangential velocity is negligible for the two cases. Özger et al. [39] measured the maximum tangential velocity u_θ/u_∞ in the wake of a BAC wing. The measured values decline from $u_\theta/u_\infty = 0.3$ at $x/b = 0.0$ to $u_\theta/u_\infty = 0.2$ at $x/b = 0.8$, while the values of the simulation decline from 0.38 to 0.28 over the same distance.

The vortex has an axial velocity deficit u_D. The value at the inflow boundary is $u_D/u_\infty = 0.8$ for engine position “A” and $u_D/u_\infty = 0.79$ for position “C”. The deficit decreases in downstream direction and after six wing span u_D/u_∞ has grown to $u_D/u_\infty = 0.1$ (Fig. 10b). Further downstream, this value remains almost constant. The deficit decreases more rapidly than the axial deficit presented by Beninati and Marshall [3]. The initial velocity deficit of their simulation is also around $u_D/u_\infty =$

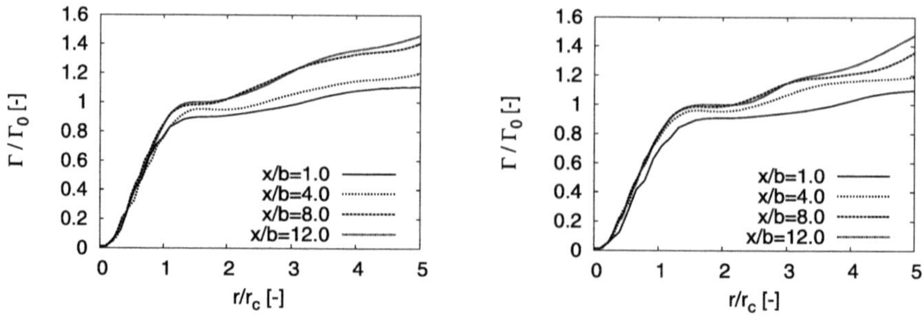

a Engine position “A” **b** Engine position “C”

Fig. 11 Circulation Γ/Γ_0 for selected positions x/b

0.8. But in contrast to the simulation the deficit remains stable up to the maximum down stream position at $x/b = 2.15$.

Fig. 10c depicts the nondimensional vorticity $(\omega_x b/2)/u_\infty$. The value decreases in downstream direction, but the difference for the two simulated cases is small.

The swirl parameter $q = u_\theta/(u_\infty - u_D)$, which is the ratio of the azimuthal to the axial velocity [24], is shown in Fig. 10d. The swirl parameter for engine position “A” is bigger than for position “C”. Note, that the values of q are quite constant for x while the values of the quantities presented above (Fig. 10a–c) are either monotonously growing or declining. For a swirl parameter $q > 1.0$, a Batchelor vortex is stable, the rotation damps all perturbations [24]. From Fig. 10d, it is obvious that the simulated vortex is stable.

From $x/b = 5$, instabilities of the vortex grow and become 3D structures. These structures lead to strong perturbations in the flow field. The analysis summarized in Figs. 10a–d was also used for the temporal development of vortical flows in [15, 26, 40]. In contrast to the results in those papers, no significant influence of the engine jet position on the wake development is observed in this study. Only the swirl parameter q in Fig. 10d shows a larger influence of the engine position. One reason could be the unphysical coupling through the periodic boundary conditions in streamwise direction in the temporal developing simulations used in the other investigations.

The circulation $\Gamma(r)$ is calculated at radius r using the vorticity ω_x

$$\Gamma(r) = \int_0^{2\pi} \int_0^r \omega_x(\zeta)\zeta \mathrm{d}\zeta \mathrm{d}\theta ,$$

where the root circulation Γ_0 is defined as the circulation at $x/b = 12.0$. In Figs. 11a and 11b, the nondimensional circulation Γ/Γ_0 is displayed for the nondimensional radius r/r_c. For a single vortex, the circulation Γ grows to a constant and maximum value. At the initial position this is Γ_0. In Figs. 11a and 11b the circulation grows beyond the maximum Γ_0, because the vorticity in the wake is also integrated at $r > r_c$, see Figs. 6 and 7 for the vorticity. The circulation reaches a local maximum at $1 < r/r_c < 2$. Then, for higher values of r/r_c, the vorticity of the wake is added to

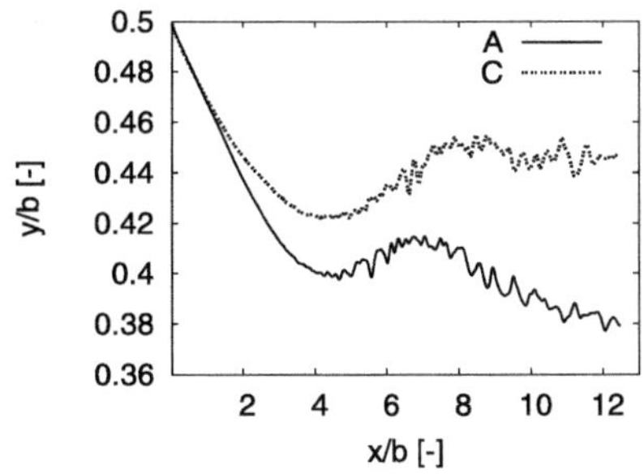

a y axis

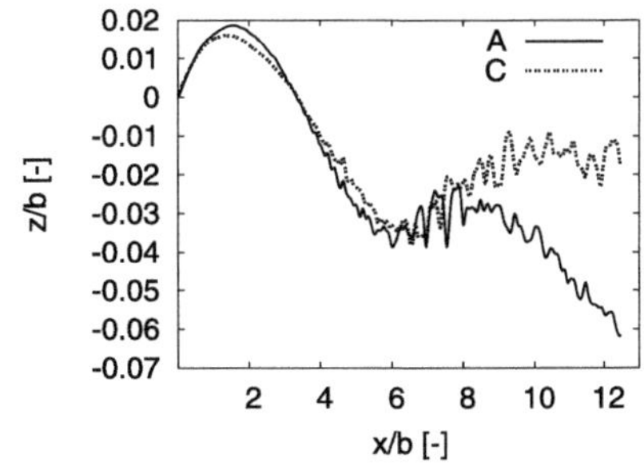

b z axis

Fig. 12 Refined position of wing tip vortex with least squares method. Comparison of bending for engine position "A" and "C"

the circulation and so the value of Γ increases. At x values downstream of $x/b \geq 8.0$, Γ rises stronger than for smaller x values, since a larger part of the wake is already wrapped around the vortex core (compare Figs. 6f and 7f).

In Fig. 12, the positions of the center of the wing tip vortex are displayed. To get a more exact position of the vortex core, the position is interpolated with an ansatz function between the discrete positions defined by the mesh. The interpolation function

$$c_0 + c_1 y + c_2 z + c_3 y^2 + c_4 z^2 = b_i \, ,$$

is of second order in both directions and calculated for the position of the vortex core and the surrounding eight points at constant x values. This leads to a linear system of equations

$$\mathbf{A}\mathbf{c} = \mathbf{b} \, ,$$

for each x position. With the nine points, nine linear equations are obtained while there are only 5 unknowns $(c_0, \ldots, c_4)$. This overdetermined system is solved with the least squares method for the vector $\mathbf{c}$. Through derivation the interpolated position of the vortex core (y_c, z_c)

$$y_c = \frac{-c_1}{2c_3} \, , \qquad z_c = \frac{-c_2}{2c_4}$$

is obtained. Figure 12a depicts the deflection on the y-axis and Fig. 12b on the z-axis. For engine position "A" — the one towards the wing tip — the deflection is larger than for position "C". At the local maximum at $x/b = 4.3$, the difference between the two trajectories is $\Delta y = 0.023b$. The differences in z-direction are relative to the deflection even larger. For case "A" the vortex looses height from $x/b = 8$, while at position "C", the vortex position is still moving upwards.

One possible explanation for the different wing tip vortex positions is the interaction of the vortex with the shear layer and the jet of the wake. At position "A", the jet is rolled up at a shorter distance than at engine position "C". Therefore, the moment needed for the rolling up and wrapping of the jet is smaller. Another explanation is

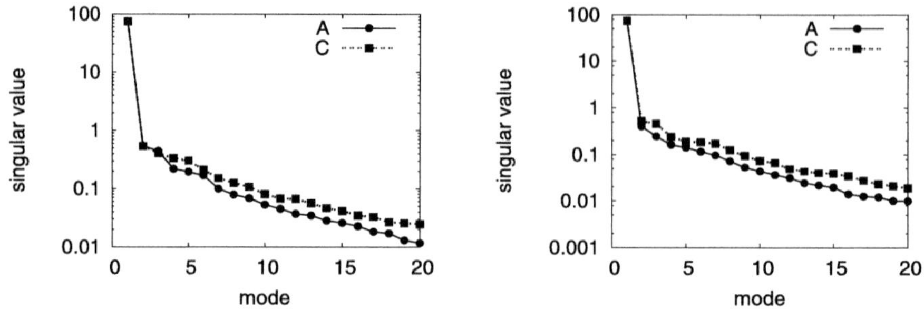

Fig. 13 Singular values of the modes 1 to 20 computed with singular value decomposition (SVD). Engine position "A" and "C". Left: $x/b = 8.0$, right $x/b = 10.0$

the angle and position of the interaction of the vortex and jet. For "C", the angle is bigger than for "A" so the jet contains a larger momentum to move the jet towards a positive y-direction.

To get a better insight of the meandering of the vortex core, a singular value decomposition (SVD) has been carried out. The similarity between proper orthogonal decomposition (POD) [4, 13] and SVD are discussed in [31]. Two positions of the flow field at $x/b = 8.0$ and 10.0 are chosen for the analysis. The energy $E = 1/2(u^2 + v^2 + w^2)$ is computed on these positions like it was computed for the FFT previously, for Figs. 8 and 9. The SVD is carried out at engine position "A" for 655 time steps and at position "C" for 644 time steps with a $\Delta t = 0.01$.

The singular values, the amplitude of the modes [1], are displayed in Fig. 13. The second and the third mode at position $x/b = 8.0$ have a similar level and are larger than the values of the higher modes. The amplitude of the higher modes of engine position "A" decrease faster than the amplitude of engine position "C". At $x/b = 10.0$, the amplitude of the second and third mode of engine position "A" decreases, while it stays for engine position "C" on a value comparable with the position for $x/b = 8.0$.

These modes are associated with the mean field for the first mode, displacement modes for the second and third mode, and higher-order perturbations for the higher modes. The reconstructed energy E of the first six modes is displayed in Figs. 14 and 15 for $x/b = 8.0$ and in Figs. 16 and 17 for $x/b = 10.0$. The structure of the reconstructed energy for the different modes of the two engine positions is similar, the largest difference can be found in mode five, see Figs. 14 and 15. Most of the energy of the vortex of engine position "A" is contained in the lower modes causing a displacement of the vortex, while at engine position "C" the energy is distributed among more higher modes.

To perform an FFT of the vortex core position, the position was translated towards the origin and then transformed from Cartesian (x, y, z) to cylindrical (r, θ, z) coordinates. The FFT was performed for 40 time steps of the unsteady flow field with constant $\Delta t = 0.1$ in a range of $5.0 \leq x/b \leq 8$ and afterwards averaged. The Fourier modes for the radius r and the angle θ are displayed in Fig. 18 for the two engine positions. The Fourier mode of r and θ has a peek at mode 6. This Fourier mode

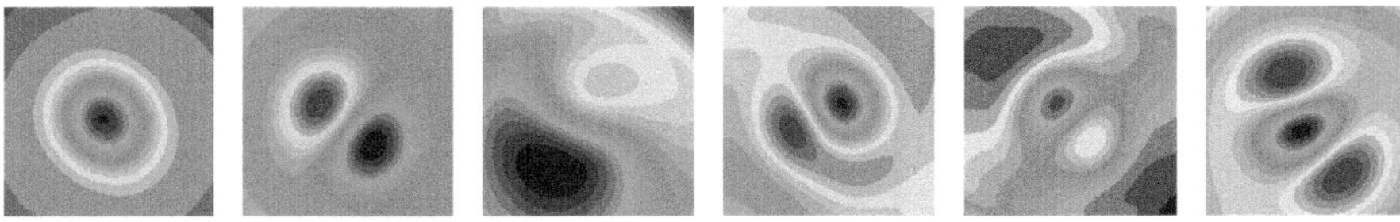

Fig. 14 Singular value decomposition (SVD) of energy at $x/b = 8.0$ for engine position "A". Modes 1 to 6

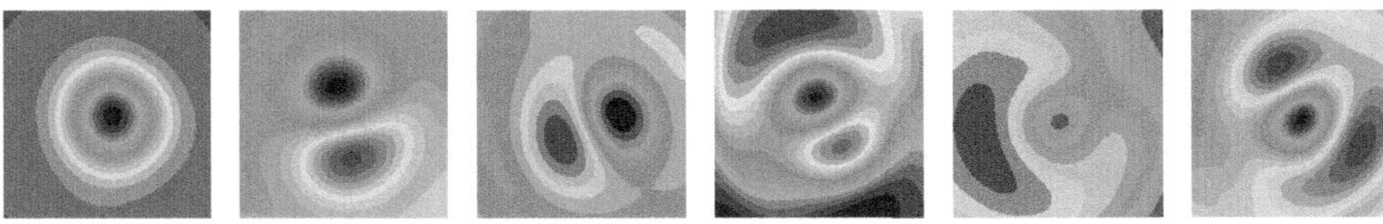

Fig. 15 Singular value decomposition (SVD) of energy at $x/b = 8.0$ for engine position "C". Modes 1 to 6

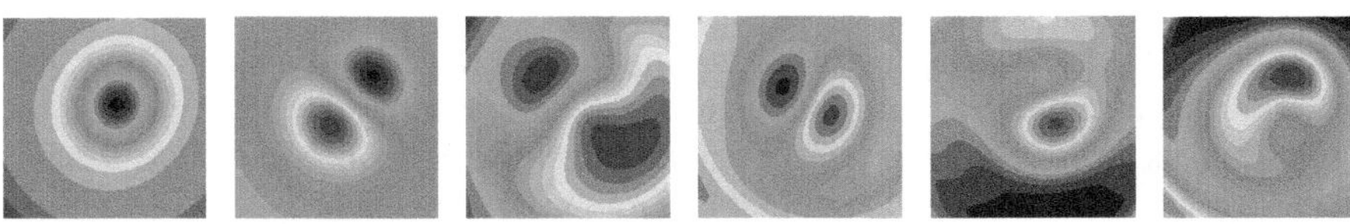

Fig. 16 Singular value decomposition (SVD) of energy at $x/b = 10.0$ for engine position "A". Modes 1 to 6

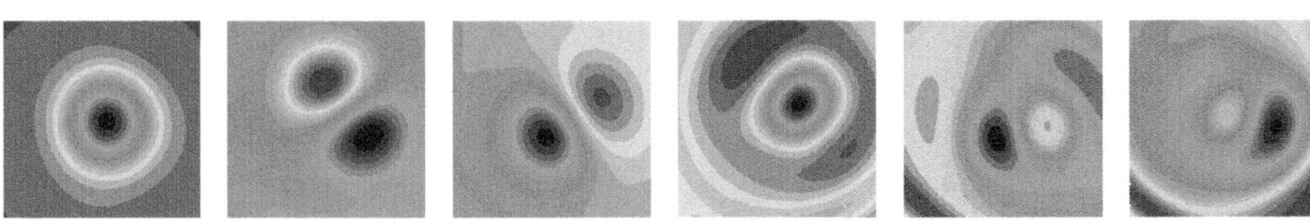

Fig. 17 Singular value decomposition (SVD) of energy at $x/b = 10.0$ for engine position "C". Modes 1 to 6

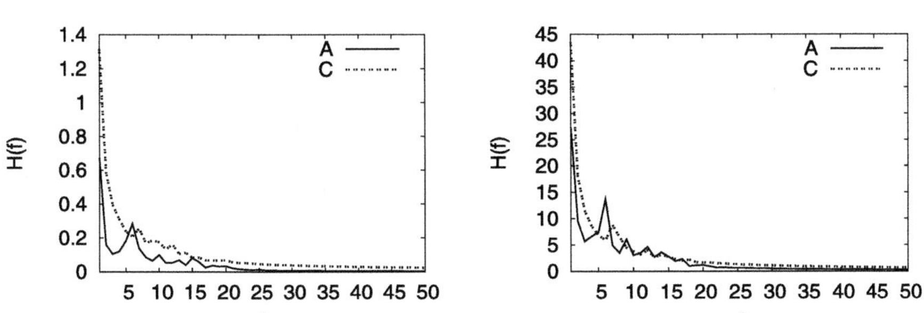

Fig. 18 Averaged Fourier modes $H(f)$ of the frequencies f of vortex position for engine position "A" and "C". Left: FFT of radius r, right: FFT of angle θ

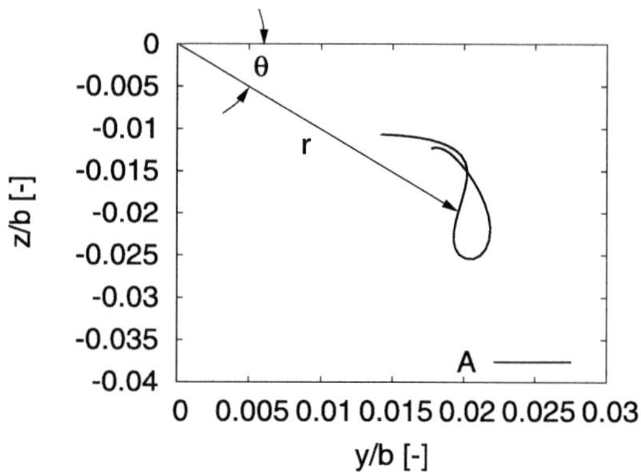

Fig. 19 Closeup of the vortex position (see Fig. 12a) for engine position "A". Displayed is z/b over y/b for $7.35 < x/b < 7.85$. The transformed variables r and θ are also sketched

corresponds to a wavelength $\lambda = 0.5b$. The vortex trajectory for a similar wavelength is displayed in Fig. 19 for the y/b and z/b direction, and $7.35 < x/b < 7.85$. Following the vortex position, both, r and θ, are changing, but the angle θ has a higher variation and causes therefore a larger Fourier coefficient.

The spatial analysis of the vortex core trajectory shows larger values of the Fourier coefficients for engine position "A" than for position "C". These modes represent the deformation of the vortex core through vortex meandering. The higher influence of the jet on the vortex meandering for engine position "A" is consistent with the findings in Figs. 4 and 5.

5 Computational Resources

All simulations were carried out on the NEC SX-8 installed at the High Performance Computing Center Stuttgart (HLRS). The results presented in Sect. 4 were computed on a domain of integration which is divided into 24 blocks, while each block resides on a single CPU. Data between the blocks is exchanged via MPI (Message Passing Interface). One node NEC SX-8 is equipped with 8 CPUs, so three nodes are used for the simulation. The workload for each CPU, that is mesh points per block, is high, but since the scheduling system prefers jobs with fewer nodes, the domain decomposition was adapted to a rather small number of processors to achieve a small turnaround time and not to a massively parallelized simulation. In principle, other distributions of the mesh points are possible.

Here, we present additionally computing statistics of another simulation which is not discussed elsewhere in this paper. This case "I" contains 40 million mesh points, that are distributed equally to 16 blocks on two nodes.

Table 1 summarizes the performance of the two different mesh topologies. For a realistic simulation of the wake, the whole wake with both wing tip vortices must

Table 1 Sample performance on NEC SX-8

	Case "A"/"C"	Case "I"
Number of CPU	24	16
Number of Nodes	3	2
Mesh size	$63,2 \cdot 10^6$	$40.2 \cdot 10^6$
Mesh points/CPU	$2.7 \cdot 10^6$	$2.6 \cdot 10^6$
Avg. User Time [s]	44500	43100
Avg. Vector Time [s]	41735	41100
Vector Operations Ratio [%]	99.5	99.6
Avg. Vector Length	240.0	242.3
Avg. MFlop/s / CPU	5035	5500
Max. MFlop/s / CPU	5090	5550
Memory/CPU [MB]	1750	1666
total GFlop/s	120.8	88.1
total Memory [GB]	41.2	26.0

be computed. A first estimation leads to a six times larger number of mesh points for the same simulation length downstream ($12.68\,b$). For the simulation of the long wave Crow instability [8], the integration domain must be larger than 100 spans since this slowly growing instability causes a breakdown of the vortex system [41] beyond this length. For this simulation more than 3 billion mesh points would be needed with approximately 2.5 TB RAM.

6 Conclusions

A successful LES of a wake-jet interaction in the wake of a wing has been presented in this paper. The inflow boundary condition for the spatial simulation of the wake has been determined from PIV measurements in the wake of a wing with an engine mounted in two different positions. The simulation has shown that the vorticity distribution in the wake is influenced by the engine position, the Fourier coefficients of the Fourier modes $k = 2$ and 4 have depicted a different growth rates. The temporal development of the Fourier coefficients have revealed the vortex instabilities to be convected downstream. While the tangential velocity, the axial velocity deficit, and the streamwise vorticity component of the tip vortex have shown no significant difference for the two engine positions, the swirl parameter showed different values.

As shown, the main difference for the two cases is the trajectory of the vortex core. Engine position "C" has generated a larger deflection of the vortex core than the position "A".

Vortex meandering analyzed by the singular value decomposition (SVD) has shown a similar behavior for the two engine positions. The Fourier transformation revealed for the engine position "C" higher Fourier modes than for engine position "A".

Further investigation of the wake-jet interaction will focus on different cases like a fully turned off engine and also high lift devices like flaps, which produce additional vortices. Furthermore, a simulation of the wake of the full wing with both wing tip vortices is required to predict the growth of instabilities in the near and extended near field.

Acknowledgements The support of this research by the Deutsche Forschungsgemeinschaft (DFG) in the frame of SFB 401 is gratefully acknowledged.

References

1. Ahmed, S., Raju, D., Chaturvedi, S., Jha, R.: Modal analysis for a bounded-wave EMP simulator — Part I: effect of test object. IEEE Transactions on Electromagnetic Compatibility **47**(1), 171–182 (2005)
2. Baker, G.R., Barker, S.J., Bofah, K.K., Saffman, P.G.: Laser anemometer measurements of trailing vortices in water. J. Fluid Mech. **65**, 325–336 (1974)
3. Beninati, M.L., Marshall, J.S.: An experimental study of the effect of free-stream turbulence on a trailing vortex. Experiments in Fluids **38**, 244–257 (2005). DOI 10.1007/s00348-004-0904-1ELECTR:2
4. Berkooz, G., Holmes, P., Lumley, J.L.: The proper orthogonal decomposition in the analysis of turbulent flows. Annual Review of Fluid Mechanics **25**, 539–575 (1993)
5. Bristol, R.L., Ortega, J.M., Marcus, P.S., Savas, Ö.: On cooperative instabilities of parallel vortex pairs. J. Fluid Mech. **517**, 331–358 (2004)
6. Brunet, S., Garnier, F., Sagaut, P.: Crow instability effects on the exhaust plume mixing and condensation. In: Third International Workshop on Vortex Flows and Related Numerical Methods, *European Series in Applied and Industrial Mathematics*, vol. 7, pp. 69–79 (1999). URL http://www.emath.fr/proc/Vol.7/
7. Crouch, J.D.: Instability and transient growth for two trailing-vortex pairs. J. Fluid Mech. **350**, 311–330 (1997)
8. Crow, S.C.: Stability theory for a pair of trailing vortices. AIAA J. **8**, 2172–2179 (1970)
9. Devenport, W.J., Rife, M.C., Liapis, S.I., Follin, G.J.: The structure and development of a wing-tip vortex. J. Fluid Mech. **312**, 67–106 (1996)
10. Fabre, D., Jacquin, L.: Stability of a four-vortex aircraft wake model. Phys. Fluids **12**(10), 2438–2443 (2000)
11. Fabre, D., Jacquin, L., Loof, A.: Optimal perturbations in a four-vortex aircraft wake in counter-rotating configuration. J. Fluid Mech. **451**, 319–328 (2002)
12. Fares, E., Schröder, W.: Analysis of wakes and wake-jet interaction. In: Notes on Numerical Fluid Mechanics, vol. 84, pp. 57–84 (2003)
13. Feeny, B.F., Kappagantu, R.: On the physical interpretation of proper orthogonal modes in vibrations. Journal of Sound and Vibration **211**(4), 607–616 (1998)
14. Gago, C.F., Brunet, S., Garnier, F.: Numerical Investigation of Turbulent Mixing in a Jet/Wake Vortex Interaction. AIAA J. **40**(2), 276–284 (2002)
15. Gallaire, F., Chomaz, J.M.: Mode selection in swirling jet experiments: a linear stability analysis. J. Fluid Mech. **494**, 223–253 (2003)
16. Gerz, T., Holzäpfel, F.: Wing-tip vortices, turbulence, and the distribution of emissions. AIAA J. **37**(10), 1270–1276 (1999)
17. Ghosal, S., Moin, P.: The basic equations for the large eddy simulation of turbulent flows in complex geometry. J. Comput. Phys. **118**, 24–37 (1995)
18. Holzäpfel, F., Gerz, T., Baumann, R.: The turbulent decay of trailing vortex pairs in stably stratified environments. Aerosp. Sci. Technol. **5**, 95–108 (2001)

19. Holzäpfel, F., Hofbauer, T., Darracq, D., Moet, H., Garnier, F., Gago, C.F.: Wake vortex evolution and decay mechanisms in the atmosphere. In: Proceedings of 3rd ONERA–DLR Aerospace Symposium, p. 10. Paris, France (2001)
20. Holzäpfel, F., Hofbauer, T., Darracq, D., Moet, H., Garnier, F., Gago, C.F.: Analysis of wake vortex decay mechanisms in the atmosphere. Aerosp. Sci. Technol. **7**, 263–275 (2003)
21. Huppertz, G., Klaas, M., Schröder, W.: Engine jet/vortex interaction in the near wake of an airfoil. In: AIAA 36th Fluid Dynamics Conference. San Francisco, CA, U.S.A (2006). AIAA-Paper 2006-3747
22. Huppertz, G., Schröder, W.: Vortex/engine jet interaction in the near wake of a swept wing. In: 77th Annual Meeting of the GAMM. Berlin, Germany (2006)
23. Jacquin, L., Fabre, D., Sipp, D., Theofilis, V., Vollmers, H.: Instability and unsteadiness of aircraft wake vortices. Aerosp. Sci. Technol. **7**, 577–593 (2003)
24. Jacquin, L., Pantano, C.: On the persistence of trailing vortices. J. Fluid Mech. **471**, 159–168 (2002)
25. Jeong, J., Hussain, F.: On the identification of a vortex. J. Fluid Mech. **285**, 69–94 (1995)
26. Labbe, O., Maglaras, E., Garnier, F.: Large-eddy simulation of a turbulent jet and wake vortex interaction. Comput. & Fluids **36**(4), 772–785 (2007)
27. Laporte, F., Corjon, A.: Direct numerical simulations of the elliptic instability of a vortex pair. Phys. Fluids **12**(5), 1016–1031 (2000)
28. Laporte, F., Leweke, T.: Elliptic Instability of Counter-Rotating Vortices: Experiment and Direct Numerical Simulation. AIAA J. **40**(12), 2483–2494 (2002)
29. Le Dizès, S., Laporte, F.: Theoretical predictions for the elliptical instability in a two-vortex flow. J. Fluid Mech. **471**, 169–201 (2002)
30. Leweke, T., Williamson, C.H.K.: Cooperative elliptic instability of a vortex pair. J. Fluid Mech. **360**, 85–119 (1998)
31. Liang, Y.C., Lee, H.P., Lim, S.P., Lin, W.Z., Lee, K.H., Wu, C.G.: Proper orthogonal decomposition and its applications — Part I: Theory. Journal of Sound Vibration **252**, 527–544 (2002)
32. Loiseleux, T., Chomaz, J.M., Huerre, P.: The effect of swirl on jets and wakes: Linear instability of the rankine vortex with axial flow. Physics of Fluids **10**(5), 1120–1134 (1998)
33. Lu, G., Lele, S.K.: Inviscid instability of compressible swirling mixing layers. Physics of Fluids **11**(2), 450–461 (1999)
34. Meinke, M., Schulz, C., Rister, T.: LES of spatially developing jets. In: Computation and Visualization of Three-Dimensional Vortical and Turbulent Flows, Notes on Numerical Fluid Mechanics. Vieweg Verlag (1997)
35. Meunier, P., Dizes, S.L., Leweke, T.: Physics of vortex merging. C. R. Physique **6**, 431–450 (2005). DOI http://dx.doi.org/10.1016/j.crhy.2005.06.003
36. Meunier, P., Leweke, T.: Three-dimensional instability during vortex merging. Phys. Fluids **13**(10), 2747–2750 (2001)
37. Meunier, P., Leweke, T.: Elliptic instability of a co-rotating vortex pair. J. Fluid Mech. **533**, 124–159 (2005)
38. Miake-Lye, R., Martinez-Sanchez, M., Brown, R.C., Kolb, C.E.: Plume and wake dynamics, mixing, and chemistry behind a high speed civil transport aircraft. J. Aircraft **30**(4), 467–479 (1993)
39. Özger, E., Schell, I., Jacob, D.: On the Structure and Attenuation of an Aircraft Wake. J. Aircraft **38**(5), 878–887 (2001)
40. Paoli, R., Laporte, F., Cuenot, B., Poinsot, T.: Dynamics and mixing in jet/vortex interactions. Phys. Fluids **15**(7), 1843–1860 (2003). DOI 10.1063/1.1575232
41. Saudreau, M., Moet, H.: Characterization of Extended Near-Field and Crow Instability in the Far-Field of a Realistic Aircraft Wake. In: P. Neittaanmäki, T. Rossi, S. Korotov, E. Onate, J. Periaux, D. Knörzer (eds.) European Congress on Computational Methods in Applied Sciences and Engineering ECCOMAS 2004 (2004)
42. Schlichting, H., Truckenbrodt, E.A.: Aerodynamik des Flugzeuges, vol. 2. Springer, Berlin, Heidelberg, New York, Barcelona, Hongkong, London, Mailand, Paris, Singapur, Tokio (2001)

43. Sipp, D.: Weakly nonlinear saturation of short-wave instabilities in a strained Lamb-Oseen vortex. Phys. Fluids **12**(7), 1715–1729 (2000)
44. Spalart, P.R.: Airplane trailing vortices. Annual Review of Fluid Mechanics **30**(1), 107–138 (1998)
45. Stumpf, E.: Untersuchung von 4-Wirbelsystemen zur Minimierung von Wirbelschleppen und ihre Realisierung an Transportflugzeugen. Ph.D. thesis, Aerodyn. Inst., RWTH Aachen (2003)
46. Stumpf, E., Wild, J., Dafa'Alla, A.A., Meese, E.A.: Numerical simulations of the wake vortex near field of high-lift configurations. In: P. Neittaanmäki, T. Rossi, S. Korotov, E. Oñate, J. Périaux, D. Knörzer (eds.) European Congress on Computational Methods in Applied Sciences an Engineering ECCOMAS 2004. Jyväskylä (2004)
47. Waleffe, F.: On the three-dimensional instability of strained vortices. Phys. Fluids **2**(1), 76–80 (1990)
48. Yin, X.Y., Sun, D.J., Wei, M.J., Wu, J.Z.: Absolute and convective instability character of slender viscous vortices. Phys. Fluids **12**, 1062–1072 (2000). DOI 10.1063/1.870361
49. Zurheide, F., Schröder, W.: Numerical analysis of wing vortices. In: C. Tropea, S. Jakirlic, H.J. Heinemann, R. Henke, H. Hönlinger (eds.) New Results in Numerical and Experimental Fluid Mechanics VI, *Notes on Numerical Fluid Mechanics and Multidisciplinary Design*, vol. 96, pp. 17–25. Springer Berlin, Heidelberg, New York (2008)

GPSR Compliance

The European Union's (EU) General Product Safety Regulation (GPSR) is a set of rules that requires consumer products to be safe and our obligations to ensure this.

If you have any concerns about our products, you can contact us on ProductSafety@springernature.com

In case Publisher is established outside the EU, the EU authorized representative is:

Springer Nature Customer Service Center GmbH
Europaplatz 3
69115 Heidelberg, Germany

Batch number: 10620141

Printed by Printforce, the Netherlands